How to Think With AI

A Simple Guide to Boost Brain Power,
Creativity, and Performance

Alison McCauley

COLLINS & CROSS
PUBLISHING

Published 2025
Printed in the United States of America

ISBN: 978-0-9980420-6-0 (print), 978-0-9980420-9-1 (epub)

Library of Congress Cataloging-in-Publication Data
Control Number: 2024926460
Name: Alison McCauley
Title: How To Think With AI

Ordering information: discounts are available on quantity purchases by corporations, associations, and others. For details, write to orders@unblockedcollective.com.

COLLINS & CROSS
PUBLISHING

To the explorers who never stop asking "what if,"
inventors who see solutions in shadows,
and pioneers translating discoveries into hope.

To the leaders forging bridges across our divides,
and the people who bring neighborhoods to life.

To the innovators expanding the frontier of human wellness,
and practitioners turning knowledge into healing.

To the parents cultivating curious minds and brave hearts,
mentors whose insight ignites countless journeys,
and those with generous hearts who dedicate
themselves to lifting others.

To all who dare to reimagine what could be.

May we be brave enough to define new horizons,
wise enough to build bridges to that future,
and determined enough to make it real.

Contents

Part Three: Thinking 2.0

Afterword
Notes

Introduction
Powering Up

1

From Eye Rolls to Aha: The Day AI Became Real for My Family

My teenage daughter turned to me, her eyes wide. "Mom," she said with intensity, "this. . . this is going to change everything." In that instant, I watched the months I'd spent struggling to convey the impact of today's artificial intelligence suddenly crystallize. Her face lit up with realization. This was the moment I'd been striving for.

The year before, on November 30, 2022, OpenAI gave the world ChatGPT, and suddenly anyone with internet access could tap into sophisticated artificial intelligence (AI) as easily as typing into a search engine. The reaction surprised even OpenAI. The company's chief scientist, Ilya Sutskever, told *MIT Technology Review*, "When we made ChatGPT, I didn't know if it was any good. When you asked it a factual question, it gave you a wrong answer. I thought it was going to be so unimpressive that people would say, 'Why are you doing this? This is so boring!'"[1]

However, wrapping a familiar, easy-to-use interface around an advanced AI model trained on much of the internet's knowledge proved irresistible to the public. Within five days, one million people had firsthand experience with how far AI had come. In two months, ChatGPT had reached 100 million users to become the most successful consumer software launch in history (Facebook took five years to reach that milestone) and had set off a new era of AI acceleration.

Meanwhile, in my house, I was battling to persuade my family to "get" why it mattered. I gathered them around the hearth one evening, hot chocolate in hand, to throw story challenges at ChatGPT—and it filled the screen with intricate tales typed out at superhuman speed. Another evening I showed them how to generate images instantly with a line of text, and my daughter demanded fluffier and fluffier puppies until fluff consumed every pixel. They smiled. They were amused. But no one was grasping the magnitude of this moment.

Perhaps my most ill-fated attempt was when I switched an empty plate at the dinner table for a computer monitor and told my family that our dinner guest that night would be the bot I'd built to facilitate a discussion about artificial intelligence. One kid scoffed, loudly. The other rolled her eyes. My dad watched me fuss over my setup with thinly veiled concern. As a mom who works in AI, my family was used to me nerding out over dinner—but this, they told me, went too far. "You're giving me secondhand embarrassment," my son groaned.

I had failed. They weren't *really* seeing the power of this technology. They weren't grasping the immense impact it would have. They were still thinking of AI—a force that I believe will shape their future—as a toy.

Finding Value on Your Own Terms

Then came a day that changed everything. My daughter approached me with a problem—one that would finally bridge our AI divide.

She was getting crushed by the hundreds of pages of academic papers on political theory that she needed to understand. An exceptionally smart girl with a processing disorder, she had perfected her own approach to learning material in a way that worked for *her* brain: thoughtful study plans accompanied by gorgeous notebooks bursting with colors, drawings, and pages of neat handwriting. But the academic papers were dense, the topic unfamiliar—and they were breaking all the methods she had honed over the last decade.

"Honey," I said, "what about thinking with AI?"

After an intense deep dive into the responsible use of AI (this time she was listening), we sat down to create a personalized learning tool that could pull the key points from those hundreds of pages into a table, complete with explanations on two levels of sophistication and page references to easily locate the source material. And then she tested it. In under a minute, her bot (which she named Bob) generated a map to help her navigate the work and comprehend the hundreds of pages of concentrated text. As she explored each point, testing whether she agreed with the output along the way, she found that she could ingest the concepts with a nuance and understanding that had been previously hidden from her view. Finally, she was armed with a solid comprehension of the material. Now she had a foundation of knowledge she could build on, writing her own perspective and insights in her notebook with brightly colored pens. "It's like it unlocked my brain," she explained to me later.

Over that single minute, as my daughter watched Bob fill the screen with what seemed like magic, she grasped something far deeper than during my many months of coaxing. She had gotten a glimpse that AI could help her to elevate her own thinking.

As I watched this unfold, I was hit by a bolt of clarity. I realized all my attempts to show my family the value and potential impact of AI had been misguided. They didn't need abstract demonstrations. Instead, my family needed to understand how to connect the technology to their own pressing challenges.

This moment sparked my new mission to develop a simple approach to help people directly connect AI capabilities to their own needs and life challenges. In other words, to help anyone to discover meaningful value from the technology we now have available. Over the last year, I immersed myself in an intensive project, conducting research and real-world testing to craft and perfect a straightforward yet powerful methodology so anyone could learn how to Think with AI.

I share the results of this work here in these pages and on my website, ThinkwithAI.org. Discovering how to get AI to work for you is a deeply personal journey, and it's unique for everyone. As you embark on this journey to Think with AI, I'm honored to be your guide.

2
Why I Wrote This Book

We are standing at the edge of the greatest cognitive revolution in history. AI pioneers have unlocked a new superpower, and it's redefining the boundaries of human potential.

So far, the results are uneven.

Some people have already discovered how to weave these tools into their daily lives, using them to test ideas, expand their thinking, and tackle complex challenges that once seemed insurmountable. They'll tell you AI is a wildly powerful technology that helps them create and perform at a level that was impossible just a year ago. They'll say it's an imaginative, creative thought partner—that it's made them a better manager, thinker, entrepreneur, learner, parent, or partner. That they can't imagine life without it.

Far more people struggle to find meaningful ways to wield AI.

Perhaps they have dabbled a bit, getting an interesting or amusing result, or just as likely something totally underwhelming. Some may have found it useful for a one-off task—such as drafting a difficult letter or planning a party menu—but AI simply didn't "stick." Even with AI just a browser tab away, they don't reach for it. Its real-world, everyday value remains elusive.

Countless brilliant minds—executives, educators, entrepreneurs, and leaders across every sector—are missing the opportunity to amplify their impact with AI.

I wrote this book to close this gap.

I was concerned by the sheer number of people I encounter who are still on the sidelines as the world starts to shift under their feet. As AI-enhanced people drive a new baseline, suddenly everyone else will be operating at a fraction of their potential. It's not because they are less capable but because the ceiling of what's possible has risen.

This is why I spent the last year studying and codifying the behavior of early adopters who extract value from AI daily across their personal and professional lives. I captured the secrets of how they learned to Think with AI to achieve what was impossible on their own.

I've transformed these insights into a structured learning journey that will boost your personal AI Quotient, or AIQ[i], and help you succeed on the new AI-powered playing field we all face today.

Your personal AIQ is a measurement of your ability to enhance your creativity and performance by thinking with AI.

Embracing a Human-Centered Approach

This book, and the learning journey I will share, focus on a specific facet of AI: harnessing the technology to boost our own cognitive performance. While AI is being deployed across all kinds of use cases where it makes decisions and acts on them autonomously without human intervention, our focus here is different. We're exploring how we—as people—can increase our cognitive capabilities and capacity to create by *collaborating with AI.*

---

i   The term AIQ was originally used to refer to AI's abilities but has more recently been used to refer to an organization's readiness and capability to leverage AI. Several analysts and consulting firms have developed frameworks to assess an organization's AIQ, which involves factors from data readiness to governance and ethics processes. What we are talking about here is much more straightforward: personal AIQ is a measurement about how well you leverage AI to boost your personal performance.

Even if your organization ultimately aims for more comprehensive automation, the approach outlined here serves as an excellent starting point to help your teams familiarize themselves with AI's strengths, limitations, and capabilities. The rapid advancement of "AI agents" (these are systems designed to make and execute decisions independently) marks an exciting new frontier in automation. However, given our still-maturing understanding of how best to use the technology, many AI use cases still rely on what's called "human-in-the-loop." This means that AI informs decision-making, but people remain in charge of interpreting results and driving actions, rather than trusting the system to act on its own.

Thinking with AI inherently emphasizes human-in-the-loop: we are using AI as a cognitive enhancer—to help *us* think smarter and better—rather than as a replacement. This approach offers an ideal way to jump in now, get firsthand experience with the technology, and build AI muscle across an organization. In fact, as AI reshapes the competitive landscape, forward-thinking organizations are discovering that success lies not just in implementing AI systems, but in helping their teams to Think with AI. Ensuring everyone on your team has direct experience with AI's capabilities is crucial for any organization hoping to fully leverage this technology—and this is why I consistently advise pairing AI automation initiatives with efforts to help people throughout your organization learn to Think with AI.

A Guide to Getting AI to Serve You

Discovering how to get AI to serve *you*—starting with that breakthrough "aha" moment when you first see AI deliver something truly useful—is a personal journey that looks different for everyone. A CEO discovered how to use AI as a sounding board for strategy. A 94-year-old found she could use AI to recall words she was struggling to remember. A marketing leader learned how to use AI to accelerate campaign development. A non-profit director discovered how to apply for grants in a fraction of the time it took before. And a mom of four found she could use AI to help her think through strategies for supporting her kids with the challenges they faced both in the classroom and in the schoolyard. With each discovery, I would hear the same realization: "I didn't know it could do that!"

This book serves as a guide for this personal journey. I'll help you navigate through the noise, center you on what's happening in this moment, and provide a roadmap to help you match your unique needs and the new capabilities that AI offers you to discover how AI can be useful in *your* life. In this journey, we will also cover:

How to shift your mindset to get more out of AI. AI works differently from any other software we've experienced. To get value from these new tools, we need to first understand a new way of interacting with software.

Advice to move past procrastination. Procrastination often gets in the way of diving into AI. I not only share why you should feel an urgency to learn more about these new tools but also offer practical strategies to jump into a successful learning journey.

Curated recommendations to help you take your learning further. I've curated a library of online resources to help you plug into a community of shared learning.

The Next Digital Divide: Which Side Will You Be On?

My hope is that this book will help push back against a new kind of inequality that is quietly taking shape around us.

Early adopters are positioning themselves and their organizations for an AI-infused future by diving in to build AIQ even as the technology and our understanding of how to use it are in their infancy. Yet for many people, AI remains an abstraction—a tool seemingly reserved for other people—tech enthusiasts or specialists—too mysterious or out of reach to be useful in their daily lives or work. From business professionals striving to grow their companies, to social entrepreneurs stretching limited resources to build nonprofit organizations, to small business owners driving local economies, to parents balancing career and family life, many people who play vital roles in our society and economy risk missing out on the potential to supercharge their work.

For some, it is a matter of perception; for others, the demands of daily life have simply pushed AI exploration to the back burner. But as this new digital divide widens, it threatens to reshape the landscape of personal and professional success.

It is not too late to start this journey today, but soon it will be. This genie is not going back in the bottle. In fact, as you will learn in the chapters ahead, AI-driven change is only going to accelerate.

The technology is already intertwining with our lives, weaving its way into the digital experiences we have every day. Chatbots pop up to offer AI help as we go about our daily business at work. Its images infiltrate our social media feeds. Seemingly every day, there is a new headline about it. And any one of us can tap right into the raw power of some of the most advanced AI models ever made by just opening a tab on our browser.

I have spent three decades at the front edge of technology

revolutions, from the birth of the dotcom era to the rise of mobile computing. I have devoted my career to understanding how emerging technology impacts people—individuals, organizations, cultures, and society. My work has given me a close-up view of the challenges that come from innovation that has accelerated to a pace that people and institutions—our school systems, the organizations we work for, and governments—struggle to keep up with. I have witnessed firsthand how each wave of change creates a divide between those who adapt quickly and those left behind. How even as these technologies become integral to our lives, their benefits are unevenly distributed.

Now, with AI, we are facing perhaps the most significant technology shift of our lifetimes—one that promises to reshape not just how we work but what is even humanly possible.

The moment OpenAI opened the gates to reveal ChatGPT, everyone—you, your neighbors, kids, community, and coworkers—had the opportunity to tap into advanced AI. And it brought us closer to the precipice of the next digital divide.

Two core motivations fuel my mission to help people Think with AI. First, I have witnessed how those who skillfully and ethically harness the immense potential of these tools can achieve new levels of performance and creativity—and I believe this capability should be available to all of us, not restricted to a relative few. Second, the responsible evolution of AI technology demands diverse perspectives to effectively mitigate its inherent risks—and that starts with a broader spectrum of individuals gaining substantive experience with AI, understanding both its remarkable value and its complex challenges.

My hope is to make it possible for more people to responsibly tap into this power.

Overcoming Barriers Between You and Value

AI is unlike anything we've experienced before. It is nothing like the software we've used for years. We've never had apps or websites that interact with us as AI does. And yet, there are no instruction guides or manuals. It is just thrown at us. We are expected to figure out on our own what to do with it. And believe me, this is a discovery process full of trial and error.

Grassroots communities of early adopters have sprung up everywhere online. They are passionately swapping ideas, sharing experiences, and learning together—and seemingly never sleep in their quest to keep up with the rapidly shifting world of AI.

But for most, navigating through all the noise, voices, and conflicting views in this strange and ambiguous world of AI is overwhelming, especially when we're already facing the pressing responsibilities and everyday demands of our modern lives. Even if we suspect AI could be valuable, it's just too hard to prioritize the time to dig in and figure out how to get there.

I repeatedly see three key hurdles that most commonly keep people from finding value in the technology:

They're not sure how to start. They may find the tech interesting, and may have played with it, but don't know how to transition from these early experiments to finding real value.

They're confused by extreme narratives. We are all getting caught between two polarized messages: that we need to slow AI innovation because it presents an extinction risk to humanity, and that we need to accelerate AI innovation to help us fight the wicked, complicated problems our world faces. While no one knows what will happen—the reality is likely to end up somewhere in between—when extreme messages hit the headlines, they can deter people from exploring how these tools can bring value to their daily lives.

They're overwhelmed by the speed of change. The pace of change in this space, which is stunning even for AI experts, is leading to "FONKU": Fear of Not Keeping Up. FONKU can make us feel so overwhelmed that we don't know where to jump in or we do not engage at all.

This book will help you overcome these barriers so you can stop relying on secondhand knowledge and learn how to make AI work for you. By the time you finish these pages, AI won't be some distant, intimidating technology. You'll have a powerful ally at your side—always on, ready to assist when you feel stuck or limited by your knowledge or capabilities or to propel you to new levels of creativity and performance.

Let's jump in!

3
How to Use This Book

I've designed this book with a single goal: to make your journey into AI both accessible and immediately valuable. There are three parts.

Part One, Beyond Human Cognition: Discover how AI can boost your creativity and brain power—while staying true to your human strengths (page 23).

Part Two, Decoding Change: See why AI will reshape our world—and why you can't afford to wait to tap into its capabilities (page 47).

Part Three, Thinking 2.0: Get a proven, step-by-step approach to making AI serve your unique needs (page 71).

The guidance I provide will be helpful whether you are exploring how to use AI in your personal life, in your professional life, or as a team.

Evolving Definitions: How "AI" Became Synonymous with Generative AI

Generative AI is only one kind of AI, but it quickly grabbed the spotlight to subsume the way we use the term "AI." When media or non-technical people say AI today, they are often using it as shorthand to refer to "generative AI" (also referred to as genAI). Some now refer to other areas of AI as "traditional AI," which can make technologists working in that space cringe—especially when that work is cutting edge. Generative AI produces original content, while traditional AI is primarily designed to classify, recognize patterns, and make predictions based on existing data. Many of our digital experiences have depended on traditional AI for years. When you get a recommendation from Netflix, navigate with Waze, drive a Tesla, or request an Uber, you are benefiting from traditional AI. This book focuses specifically on the capabilities of generative AI, and I am also going to use this new shorthand.

When I use the term "AI" on its own, I am speaking specifically of generative AI.

I included this chapter to give you some guidance on how to get the most out of this book. My goal to make the immensely complex world of AI accessible to more people led to three key design choices for this book:

Pair the book with a website to provide fresh updates. The speed of AI advancement demanded a unique solution: while this book provides your foundation, I've created a companion website (ThinkwithAI.org) that continues to evolve. While you will find foundational knowledge and frameworks in these pages, the website offers a curated collection of resources—not just my own, but also valuable insights and perspectives from others in the space. By making the website freely available to everyone, I hope to promote wider familiarity and understanding of the new capabilities AI tools offer us all.

Introduce, don't overwhelm with, the ethical implications of AI. In my decades of working with emerging technology, I've never encountered anything that raises more complex ethical questions than AI. However, I've designed this book to provide an accessible introduction to AI rather than an exhaustive exploration of these issues. We desperately need a broader and deeper understanding of AI's risks, implications, and downstream impacts. This book will give you a foundation for understanding, but it does not cover everything you should know about how to use AI responsibly. My hope is that you'll be inspired to dive deeper—and a great place to start is with the additional resources included on my companion website that more fully explore this critical topic.

Give you paths to fast-track this material. My priority is to help you start using AI tools right away in ways that deliver real value to you, and I encourage you to choose your own speed and depth for moving through this material. If you want to leap in right away, go straight to the three-step A³ Framework in Part Three, and pair it with the exercise sets you'll find on my website referred to in that chapter. If you want to ground yourself in what's happening in this moment, take the time to also read Part One and Part Two. If you are reading this book as part of a team, make sure you don't skip Chapter Sixteen because it will give you some helpful tips to support your team's learning journey. The entire book is designed to be easily skimmed so that you can quickly find and dip into the areas that are most interesting to you.

However you choose to engage with this material, you've already taken the most important step: deciding to better understand and harness AI's potential.

4

Above All, Use AI Responsibly

Every human can boost their capacity to create with AI. But not every human will leverage this new power for good. Like any tool since the dawn of humanity—from stone and steel to the internet—AI will be used for both positive and negative ends. We're already witnessing bad actors employing AI to amplify their harmful impacts, even as others use it to advance solutions to some of the toughest problems we collectively face.

Some uses of AI draw clear ethical lines—contrast researchers accelerating cancer research with criminals automating sophisticated fraud. But in your daily life, you'll likely face more nuanced choices and complex ethical terrain.

If AI helped spark your initial ideas, does it deserve citation? Can students ethically use it as a thought partner before writing an essay on their own? What if your coworker is using AI to produce superior work but your corporate policies on the use of AI are unclear? Should they be promoted, disciplined, or simply required to share their AI-assisted methods transparently?

We're navigating what I call the "in-between time"—a period where our traditional systems and ethical frameworks haven't caught up to this disruptive technology shift. While AI has been in development for many decades, the complicated challenges that broad access to sophisticated AI introduces are new. We

17

lack guidelines and social norms to navigate these gray areas, and both the capabilities and spread of this technology have already outpaced our human, organizational, and regulatory frameworks. AI is likely to put us through one of the most tumultuous periods of change in the labor landscape that we will face across our lifetimes. But our world also faces wicked challenges such as food insecurity, environmental degradation, mental health crises, water scarcity, cybersecurity threats, and pandemic preparedness, and we really could use some superhuman capabilities to help us solve them. What is clear is that AI is a massively complex topic.

Start Now, Even Without Complete Clarity

Every significant new wave of technology has brought unprecedented opportunity—along with terribly difficult problems we weren't fully prepared for. Even technology that we regard as quite basic today—such as electricity—brought both fear and excitement with their introduction.[i] But the power of AI—coupled with the staggering speed at which it is evolving—is already making both the opportunities it presents and the problems it introduces more pronounced than any other technology wave we've experienced. How we will navigate these and what this will mean to us years from now is not clear, and it won't be for some time. But if you wait for the dust to settle, and the future to become clear, you will have been left behind in that dust.

I work to help people use AI to boost learning, creativity, and cognitive abilities so they aren't left behind in this new era. Because the space is moving fast, that work means jumping in even as our understanding of how to use AI responsibly is still evolving. I can't give you specific guidance on all the ethical

---

i   It is a familiar pattern for us to be fearful of new technology and the implications we don't fully understand. When electricity was introduced in the 19th century it was seen as a dangerous, mysterious force that many were afraid to use. And this was not unfounded: early electrical wires and appliances were often poorly insulated and posed real fire and electrocution risks. Even U.S. President Benjamin Harrison was reportedly scared to use the light switches in the White House—leaving it to his staff to turn the lights on and off.

questions this technology raises because we haven't figured out the answers as a society, and there is much work ahead. What I can tell you is that the arrival of universally accessible AI puts a high moral obligation on everyone who uses these tools to be discerning and intentional as we consider where, when, and how to use them.

Don't Cede Authority to AI . . . Yet: Keep yourself firmly in the driver's seat. While AI can inform and enhance your choices, your judgment and oversight remain essential as we work to better understand, govern, and use these powerful systems. Do not delegate important decisions to AI or let its confidence sway you in a direction that doesn't feel right to you. I must emphasize that for students, this means limiting your use of tools to supporting your learning, not doing the work for you, but also requires you to work with your school administration to understand how they are evaluating the ethical use of AI.

Practice AI Transparency: Be upfront about where and how you involved AI in your work. Transparency is crucial to establishing trust in your work in a world in which it is becoming increasingly hard for us to understand whether something was created by human or machine.

Remain Vigilant for AI's Shortcomings: AI is notorious for "hallucinations" (making things up), biases, and flawed or incomplete outputs. It's critical to maintain a skeptical eye and avoid blindly trusting AI-generated content or conclusions. Maintain human scrutiny, fact-checking, and critical thinking to navigate these shortcomings responsibly.

Strive to Understand Downstream Impacts: With everything you do, always consider the wider and "downstream" impacts of your actions. A hallucination that goes uncaught and gets posted, for example, on your blog, could later appear in search results for someone else. It's like pouring chemicals down a drain that eventually contaminate our water supply permanently. Little by little, the impact compounds until we are all affected. Take responsibility for what you produce to ensure it is properly tested and vetted. In business settings, the stakes are even higher—AI-driven decisions, especially when baked into a product or service, can impact countless lives.

Advocate for Responsible AI: If you transition from light personal use to wider-scale rollouts that could impact others, rigorous evaluation, testing, and a deep understanding of AI's risks and limitations are paramount. In a corporate setting, consult your responsible AI team about guidelines and best practices. If your company lacks such a team, consider advocating for the formation of one.

Responsible AI Is an Ongoing Journey

This book focuses specifically on the early learning stages of using AI and is not meant to provide full guidance on responsible AI practices. While I touch on the new risks and tradeoffs of AI, to give these topics the depth and nuance they deserve I suggest starting with the additional resources I've curated for you on my website, ThinkwithAI.org. I take this approach so I can use these pages to zero in on one important angle of AI: giving everyone an accessible entry point to using AI to help them think—and create—at a higher level.

As we move beyond learning about AI's capabilities to incorporating AI more into our lives and work, I believe our responsibilities increase as well. It's crucial that we all labor to better understand and mitigate the risks and implications AI introduces to our world, thus increasing the chances that AI will ultimately enable human flourishing.

As your use of AI grows, keep researching and learning, and I encourage you to immerse yourself in broader conversations on the ethics of AI. I hope you will be inspired to lend your voice to advocate for the responsible and transparent use of AI among your peers, family, community, teams, and the organizations you are a part of.

Part One
Beyond Human Cognition

5
Why Think with AI?

Suddenly, anyone who can write or speak can easily tap into powerful new machine intelligence to boost their intellectual firepower and ability to create.

Today's AI offers us a creative, "imaginative" thought partner that can tackle open-ended goals and support us even if our thinking is still messy and raw. It "understands" our human way of communicating. For the first time, we can use software to truly *help us think*.

Meet your new "thought machine." No matter who you are, what demands you face, or what dreams and aspirations you have, you can harness the vast corpus of human knowledge these tools have ingested and channel it to help you with anything you want to do, achieve, or create—all without technical expertise or coding skills.

Just how much knowledge are we talking about? Today's most advanced AI systems can ingest eight trillion words—equivalent to reading about 100 million books . . . in a single month of training.[2] It would take a human, reading every day for eight hours, more than 180,000 years to read that many books. (That's over

2,000 lifetimes!)[i] By collaborating with AI, you can boost your own cognitive power by having the incredible computational power of these machines behind you, opening doors to new solutions and ideas you might never have reached on your own.

These tools are so potent that they're poised to reshape the landscape of personal and professional success, as well as organizational performance. Those who learn to tap into AI's power have cognitive leverage. They will be able to achieve more and perform better, while those who don't risk falling behind in an increasingly AI-augmented world.

This is why it's time to learn how to Think with AI.

Thinking with AI is not about using AI as a substitute for the work of our own brain. While AI will become increasingly capable of making and acting on decisions for us without our intervention, this book pursues a different goal. This book focuses on strategically enhancing our own cognitive abilities while keeping our uniquely human vision and creativity at the core.

Thinking with AI is collaborating with AI to amplify our human performance so we can achieve something we aren't capable of on our own.

This approach puts human ingenuity at the helm and frees our brain to focus on where it's really needed. It's about using the tools to do more of what we're uniquely good at—the things the machines *can't* do. AI plays a supporting role, helping us extract more strategic and creative thinking from our own minds. These new thought machines can work *for* us by rapidly generating new ideas, revealing unexpected angles, surfacing ideas we've overlooked, quickly testing and improving hypotheses, uncovering

---

i   Rates vary, but adult readers typically read between 200 and 300 words per minute (wpm). Reading 100 million books, each averaging 80,000 words, amounts to a total of eight trillion words. At a reading speed of 250 wpm and dedicating eight hours a day to reading, it would take a human approximately 182,648 years to read this many words. This equates to about 2,283 lifetimes, based on an average human lifespan of 80 years.

unexpected connections in complex systems, finding patterns in massive data sets, handling tedious grunt work, personalizing and accelerating our learning process, and much more.

While much has been said about AI's potential to boost productivity, we're exploring a more transformative opportunity here. This is about enhancing your *cognitive performance*. It's like having a tireless, always-ready collaborator helping you learn faster, understand more deeply, and create at a higher level. And it's just a click away—once you understand the basics of managing this new relationship.

In front of you is the opportunity to tap into this unprecedented reservoir of knowledge and processing power to take you beyond the natural limits of your human brain.

Our Magnificent, Fallible Human Brain

Your human brain is beautiful and astounding. When you raise your arm and catch a ball, your brain has processed a flood of complex visual information to calculate the ball's trajectory, speed, and position. It has not only tapped into the capabilities of specialized regions of your brain to detect orientation and motion, but has also pulled in top-down processing that leverages your prior experience and the context of this specific moment to guide and refine your perception of the ball's movement. And this all happens in a split second.

We can recognize and distinguish between thousands of faces, often with just a glimpse. Our brains even provide us with specialized neural circuits and a range of fine-tuned regions, often completely outside of our conscious awareness, that rapidly detect and process nuanced cues such as facial expressions, tone of voice, and body language to identify and respond to the emotions of others.

Our human brains are creative and imaginative powerhouses,

pulling inspiration from our lived experiences, emotions, and subjective perspectives to create or invent new things. We continually make novel connections between concepts and ideas or imagine scenarios that have no grounding in reality. This complex cognitive process fuels artistic expression, scientific breakthroughs, and the discovery of new tools and technology in a way that is unlike any machine. As author and journalist David Brooks beautifully wrote, the human mind "evolved to love and bond with others; to seek the kind of wisdom that is held in the body; to physically navigate within nature and avoid the dangers therein; to pursue goodness; to marvel at and create beauty; to seek and create meaning."[3] And it does it all using only the same amount of energy it takes to power a dim light bulb.[4]

Yet, our human brains also carry flaws and limitations. All of us carry artifacts from our ancient ancestors that function poorly in our modern world.

Cognitive biases, which lead to discrimination and irrational judgments,[5] originated as mental shortcuts to support the quick decision-making essential to basic survival in the resource-limited world of early Homo sapiens.[6] Our brains are constantly awash in chemistry—continually fluctuating hormones and neurotransmitters—that can impact our cognitive performance and ability to make smart decisions.[7] Fight-or-flight responses, helpful when facing lions, tigers, and bears, can be triggered by modern banalities such as a reckless driver cutting us off in traffic and lead to a cascade of physiological changes that impact our mental abilities.[8] Stress, anxiety, inadequate nutrition, and sleep deprivation—unfortunately prevalent in our modern world—diminish our attention span and problem-solving abilities.[9]

Each of us also carries a unique recipe of cognitive weaknesses. We've all been on our own individual and often lifelong journeys to discover strategies and approaches that work best for our unique cognitive profile. But no matter how much we've succeeded, we all have areas where we could benefit from a cognitive boost. Some of these challenges are small, but many are quite difficult to overcome. Executive functioning problems, such as difficulties with planning, organizing, and managing time are widespread. Diagnoses of attention deficit hyperactivity disorder (ADHD), which can cause difficulties with attention and focus, are increasing.[10] According to the Yale Center for Dyslexia and Creativity, an estimated one in five Americans has dyslexia.[11]

Despite these challenges and limitations, our remarkable human brains continue to adapt and evolve. And now, with the advent of a new era of AI, we have powerful new tools that can open up exciting possibilities to enhance our mental capabilities in ways we're only beginning to explore.

AI Gives Us New Tools

Thinking with AI puts our human needs and dreams at the center. We're still relying on the almost magical abilities of our extraordinary human brain to drive us forward. But we are leveraging the new machine capabilities to help us.

At the core are AI models, such as large language models (LLMs), which are mathematical systems trained on vast repositories of the world's digital data—from written language to videos, images, and charts. A handful of tech giants and startups have built highly advanced "frontier" models, including LLMs like GPT. However, we're also seeing new kinds of models emerge. These include smaller models that enable AI capabilities on your phone without network connectivity, and specialized models tailored to specific industries or functional domains.

We use software to implement and interact with these models in real-world applications. In fact, AI models are increasingly woven into our everyday software and digital experiences—you may have noticed options to write, create, search, or edit with AI popping up in your enterprise software or social media apps. But we can also access the capabilities of the most advanced models directly through chatbots provided by the companies that built them. ChatGPT, developed by OpenAI, is one such chatbot. It offers access to OpenAI's sophisticated GPT models through a simple interface that requires no technical knowledge.

Like our brains, AI models have limitations and flaws. To use AI effectively to enhance our thinking, we need to understand these limitations. I'll explain these tradeoffs throughout this book so you can make informed decisions about when and how to use AI.

Unlock a New Possible
with Machine + Human Intelligence

Thinking with AI means understanding how to bring machine and human intelligence together to achieve something that wasn't possible before. We combine these two powerful forms of intelligence. Our natural brain functioning—what scholar Michael Ignatieff describes as "a distinctively, incorrigibly human activity that is a complex combination of conscious and unconscious, rational and intuitive, logical and emotional reflection . . . so complex that neither neurologists nor philosophers have found a way to model it"[12]—works alongside AI's vast processing power to achieve what neither could do alone. It's a collaboration with the potential to enhance our performance as individuals, teams, and entire organizations.

Recent research offers a glimpse into one facet of AI's potential. A study of 100,000 people across eleven occupations in Denmark found that workers estimated that using ChatGPT could halve their working time for over thirty percent of their job tasks.[13] However, the promise of AI extends far beyond the relatively mundane benefit of time savings. Those who master the art of collaborating with AI use the technology as a force multiplier for their human capability—to help them push the boundaries of what's possible as a human. Here are some of the ways people with high AIQ can elevate their innate potential:

Amplify ideation: AI can significantly enhance our creative processes by generating a high volume of ideas or identifying angles and perspectives we haven't considered, often in mere seconds. It can function as a sort of superhuman brainstorming partner, rapidly exposing us to a broader set of concepts than we can achieve on our own, and stimulating our brains to think more deeply or holistically. AI can interpret and present information in various modes, such as transforming a dense report into an engaging infographic or creating a podcast-style audio summary of an academic paper's key points—bringing complex ideas to life and triggering new thought patterns.

Have you ever stared at a blank page, unsure where to begin? AI can solve this "blank page problem" by filling your screen with a draft or key ideas to trigger your thinking. Even if you reject most of these machine-generated ideas, the output often jumpstarts your brain. Or, if you've already developed your thesis but want a sounding board to identify gaps or logic flaws, AI is a round-the-clock partner that can evaluate, suggest, and help you test your thinking. For example, a CEO shared that she does her best thinking when she has a conversation with AI because it asks her tough questions that no one else is willing to—and this pushes her to new insights.

Bridge functional knowledge gaps: Many of us excel in specific areas but face challenges when tasks require working outside our core expertise. AI can help bridge these functional knowledge gaps, allowing us to perform competently across a broader range of activities. For instance, a brilliant engineer might struggle to describe a product to a non-technical audience, or a creative designer might find financial planning daunting. AI can provide quick, competent assistance, drafting plans, projections, or documents that can give people a "head start" in areas in which they are unfamiliar. This support is particularly valuable in roles that demand diverse skills, such as entrepreneurship, project management, and small business ownership.

Deliver new skills on demand: AI is increasingly capable of performing tasks that once required years of training and highly specialized skills, making it possible for anyone to create and innovate in new ways. If you don't know how to code, you can write a description of a website or a software feature in conversational language, and perhaps pair that with some simple sketches, and AI tools will produce working code for you. Similarly, even if you have no experience in video production or don't have video equipment, you can describe the video you want to make, and AI tools can create realistic videos. (While this functionality is still in its infancy, it is evolving fast.) Struggling to think of an image to pair with your blog? AI can generate ideas and create one for you.

Another area in which AI is making rapid advancements is translating text, audio, and video into a wide range of languages with a click or two. Tools that offer near real-time translation are even making it possible to hold a conversation between people who don't speak the same language. With increasingly accessible AI assistance, we may be able to eventually crush language barriers even without years of study.

Offload basic cognitive tasks: AI can handle low-level tasks at high speeds, freeing up our brains and time to focus on higher-level thinking. If you provide a starting point, even a jumbled list of goals and constraints, AI can organize these elements into logical steps and create a structured project plan. It can sift through hundreds of pages of documentation and handbooks to extract relevant information, helping you respond to customer inquiries more efficiently. AI can also analyze vast amounts of research data to surface trends and create comprehensive reports that "show its work," eliminating the need for manual sifting.

Companies are increasingly using AI to handle routine tasks such as summarizing documents, drafting presentations, developing meeting notes, or personalizing recommendations, enabling their employees to focus on more strategic and creative work.

Bridge cognitive gaps: We all have things we aren't good at. Regardless of where or when you struggle, AI tools can assist with tasks you find challenging, providing support tailored to your specific needs and learning style. Maybe you are comfortable with analytical challenges, but your mind goes blank when faced with a creative problem. Or creativity is your strength, but you can't seem to organize your work into a roadmap or schedule no matter how hard you try. Or you love writing, but editing your work is a challenge.

Tools designed for people with learning differences, such

as ADHD or dyslexia, are also becoming more sophisticated and accessible. And people are building their own custom AI bots in minutes without technical expertise, helping to ease the burden of individual challenges and enabling them to work more effectively.

Challenge perspectives: AI can serve as your personal devil's advocate, expanding your viewpoints and uncovering blind spots in your thinking. By simulating diverse perspectives, it helps you break free from habitual thought patterns, revealing insights and approaches you might otherwise overlook. When creating a document or tackling a complex problem, you can ask AI to role-play and review your work from different angles. For important decisions, AI can present counterarguments, pushing you to strengthen your reasoning or reconsider your stance. When preparing for negotiations, it can argue from the opposite side to help you hone your strategy.

This approach can be valuable for personal growth, creative projects, and navigating unfamiliar professional or personal territory. Even top experts in their fields find that AI challenges and encourages them to explore new angles or perspectives they hadn't considered before.

Turbocharge mental horsepower: AI can process, synthesize, and derive insights from vast volumes of data at speeds and scales that are humanly impossible. It's like the cognitive equivalent of a superhero that can leap tall buildings in a single bound, allowing us to transcend the physical constraints of our brains, as well as our very human need to take breaks to sleep and eat.

These tools can now ingest information in a range of modes, so no matter how sophisticated the topic, they can help us make sense of complex written, spoken, and even visual inputs. In fields such as finance, healthcare, and research, AI is used to analyze large datasets, identify patterns, and provide insights that drive decision-making and innovation, far surpassing what could be achieved before.

For example, Google DeepMind scientists used AI to read 200,000 scientific papers, find the 250 papers that were most relevant to their research, extract the data from these papers, and format this information for easy review—and it did all this work while the scientists took a break to eat lunch.[14]

The real opportunity in this moment in AI is discovering how our human brain and the new AI capabilities that we now have can best work *together*. While this is an individual journey, it also gives us an opportunity for collective impact. Each of us has the potential to uncover how AI can compensate for our weaknesses, help us amplify our strengths, and help us with our unique needs. But collectively we also have the opportunity to discover how AI could help us drive meaningful progress on the most urgent issues our world faces. Of course, to seize this opportunity, we must first learn how to harness AI effectively.

6

Engage with Intent:
Mindful Use of AI

The conversation was familiar. As we caught up over dinner, my friend confessed his own hesitation to Think with AI in his day-to-day work, even though his job is helping companies to streamline operations.

He writes beautifully and masterfully synthesizes complex strategic information. He's proud of this, and it has given him an edge. He was worried his work would be dumbed down or even that his skills could atrophy if he began to rely on AI. He was wary of losing his sharpness.

I told him he was right to worry. Effectively boosting our brains with AI is a balance that's both art and science.

Don't Short-Circuit Your Brain

The last thing we want to do is override our natural cognitive systems by turning to AI. If we truly want to optimize our performance, we must be thoughtful about when and where we integrate AI tools into our work. The key is to find our sweet spot where we are fully mining our own mental resources, extracting all the insights we can on our own, while strategically leveraging AI to enhance our capabilities.

As AI gets "smarter," there may be an increased risk of what Fabrizio Dell'Acqua of Harvard Business School calls "falling asleep at the wheel." Dell'Acqua found that workers using high-quality AI were more likely to follow its recommendations mindlessly, without critical evaluation, and as a result were less accurate in their work, while those with low-quality AI exerted more effort and exhibited better performance.[15]

We're also getting some evidence that if we turn to AI at the wrong time, it has a negative impact on our decision-making. Research has found that when we see AI's answer before forming our own opinion, we're more likely to accept its response—even when it's wrong —and this interaction can impact all our subsequent judgments.[16]

We need to discover when to rely on our own cognitive processes and when to leverage AI as a thinking partner and collaborator.

When used at the right moment, AI can supercharge our thinking and spark new directions of thought. How do we discover what is "right"? The first step is understanding more about how our brains process information.

Let Your Own Brain Take the Lead

Our brains naturally have powerful ways to process information, solve problems, and drive insight. Meanwhile, as author and journalist David Brooks describes, "The AI 'mind' lacks consciousness, understanding, biology, self-awareness, emotions, moral sentiments, agency, a unique worldview based on a lifetime of distinct and never to be repeated experiences."[17] So how do we use AI in a way that amplifies our natural human strengths *and* prevents us from being overly influenced by AI's blind spots?

Lean into Creative Discomfort

With challenging strategic or creative work, particularly tasks that demand novel connections or breaking new ground, we can often discover sparks of inspiration in the depths of the creative struggle.

This process can be exquisitely painful—a frustrating, maddening dance with our own thoughts, peppered with false starts and dead ends. As we push the boundaries of our thinking, we *will* stumble. Professor and author Brené Brown captures this perfectly: "There is no innovation and creativity without failure."[18] Ouch.

Yet this discomfort is a crucial element of our human creative process. Henri Matisse observed, "Don't wait for inspiration. It comes while working," to describe how engaging with the creative process, even when it feels difficult, often triggers breakthroughs.

The creative struggle isn't just an obstacle to overcome, but an integral part of the creative process itself. It's in its midst that we often make unexpected connections and ultimately produce our most innovative work. We need to take care to give our innate processes time and space to unfold before we look to machine intelligence for a boost. Introducing AI too early can influence our perception of possibilities and solutions, potentially narrowing our views. Those with high AIQ know when to resist the temptation to immediately turn to AI. To fully flex their uniquely human cognitive strength, they embrace some amount of creative struggle.

Keep Your Subconscious Mind Flowing

Other times, insights seem to materialize out of thin air. This is our human gift of subconscious processing at work—a powerful ability that can surface connections and ideas in ways even neuroscientists don't fully understand.

Scientists and scholars have been working to measure and explain the power of the unconscious mind for over a century. This is notoriously difficult—we can't really "see" unconscious processing, so this topic can be controversial. Interestingly, the debate isn't really about whether unconscious processing exists (most researchers agree that it does), but rather about how to prove it.[19] However, we have seen that brain activity related to decisions can be detected before a person becomes aware of their decision.[20] Research from cognitive neuroscientist Mark Beeman suggests that sudden insights that he calls "aha moments" are preceded by a burst of gamma-band neural activity in the brain's right hemisphere.[21]

You've likely experienced moments when an idea struck you with a bolt of insight—fully formed and profound—even when you weren't aware you were thinking. This happens because our conscious mind often becomes aware of insights only after they've been processed in the depths of our subconscious. Neuroscientist David Eagleman beautifully describes this phenomenon: "the conscious mind is not at the center of the action in the brain; instead, it is far out on a distant edge, hearing but whispers of the activity."[22] It's as if we have our *own* built-in robot working tirelessly in the background of our brains on our behalf, making connections and surfacing insights for us with minimal conscious effort on our part.

Many researchers believe that our subconscious mind also plays an important role in innovation: making connections, generating ideas, and making choices. For example, while both conscious and unconscious thought generate ideas, individuals who engage in unconscious thought are better at selecting their most creative ideas.[23]

Moreover, unconscious thought often produces more creative and unusual responses than conscious deliberation, a quality that's crucial for innovative thinking. Research has also surfaced the concept of incubation, a period in which the unconscious mind continues to work on a problem without conscious awareness, a process that can lead to creative and innovative insights.[24]

These are all reasons why when we work on tough challenges, we need to make sure we are making time and space for unconscious thinking to unfold instead of turning immediately to AI for answers.

A Challenge We Weren't Ready For: The Promise and Peril of AI in Education

AI offers exciting possibilities for personalized education at scale, but it also presents significant challenges, especially for young learners. As we prepare students for a rapidly approaching future, we must urgently address how to foster genuine learning in an AI-saturated world.

Concerns about academic integrity and skill development are valid and pressing. Students' temptation to offload work to AI is a real and complex issue. However, completely avoiding the technology isn't the answer—AI is here to stay and will likely shape students' future lives. Instead, we must teach responsible AI use and better understand how to use it to accelerate and support learning, rather than as a shortcut.

This transition period is fraught with difficulties. It places a terribly heavy burden on students to make ethical choices and on educators to navigate uncharted territory. For parents, engaging in open discussions about AI use is crucial. For students, maintaining a high standard of genuine learning is essential—your future success depends on it.

How Delegating Drudgery Can Free Your Brain

There is, however, a category of tasks that sucks up our mental energy and time without truly requiring our unique skills: rote or repetitive work. We may benefit from strategically offloading this kind of work to AI because it can reduce our cognitive burden, freeing up our mental resources for more high-level thinking and creative problem-solving. By letting AI handle "grunt work," we can focus our brains on the challenges that truly demand our insight and perspective.

"Cognitive load" is a nearly forty-year-old theory that describes how performance declines when people attempt to process too much information simultaneously. Extensive research has supported this concept over the years, but you've likely experienced it firsthand.

For instance, have you ever felt overwhelmed trying to plan a project while your computer constantly notifies you of incoming messages? Or struggled to follow a detailed conversation in a crowded, noisy place with lots of visual stimuli, such as an amusement park? Have you noticed how moving to a quiet spot or taking a break enables you to process more effectively? If so, you've experienced firsthand how cognitive load impacts your ability to process and retain information.

Early research suggests that delegating certain tasks to AI might reduce our cognitive load, freeing up mental capacity for more demanding work that still requires human intelligence. A paper from the National Library of Medicine explores how AI could alleviate the cognitive burden of healthcare workers, potentially reducing burnout and improving patient care.[25] In education, there's growing interest in how AI could cut cognitive load by providing personalized explanations, drills, and dialogues tailored to individual students' knowledge levels and learning differences.

While more research is needed,[26] many in the AI community, myself included, have experienced this effect firsthand. If you've ever wished for more hours in a day, you'll understand the impact of having AI take on tedious tasks for you. I delegate all kinds of tasks to AI—transforming whiteboard scribbles into coherent meeting summaries, creating grocery lists from weekly meal plans, organizing and formatting data into tables, helping me find the right word to express a thought, drafting presentation slides from my meeting notes, turning photos I've taken of slides at a conference into a recap for my team, and so much more. I have more energy for the work that really matters, and truly requires my brain power, and it makes my day feel more meaningful.

Boosting Brain Power Takes Discipline

In a world in which machine brains are right there, a click away, we must be very intentional about how we engage with AI to keep from inadvertently shutting off the natural power of our human brains.

This is an art, and it's one learned over time. We all need to discover our own sweet spot for when to turn to AI and when to rely exclusively on ourselves—and this balance may shift, depending on the project or task at hand.

Once you've found your sweet spot, it takes real discipline to stick to it—to resist the allure of AI when you know wrestling with a problem yourself would yield better results. I think of this like maintaining healthy habits in a world of temptations. Just as we summon the willpower to hit the gym and bypass the office donuts (most of the time) for our physical health, we must cultivate habits that support our cognitive vibrancy.

While there's no one-size-fits-all approach, I encourage you to resist the impulse to turn to AI for your most strategic and creative endeavors until you have taken a beat to pause, reflect, and wrestle with the challenge on your own first. Especially when facing a blank page or feeling "stuck" on how to take a next step, try leaning into the struggle—it can be a critical component of the creative process. Once you've pushed your own thinking to its limits, consider bringing in AI as a collaborator to enhance, refine, or expand your work.

Keeping It Human in a Digital World

As you navigate this discovery process, be mindful of which mental activities bring you genuine pleasure. There's a reason some of us bake bread despite having a great bakery nearby. Sometimes, the act itself—be it handcrafting or intellectual exploration—brings satisfaction that outweighs the need for perfection. The meaning and intellectual energy we derive from doing something ourselves can be more valuable than creating the "best" outcome. It can mean more simply because it's ours and we put in the effort to create it.

This personal, direct engagement can go further than satisfaction—on occasion, it can also spark new ideas and breakthroughs. We have a rich history of groundbreaking discoveries that came from humans simply moving through our world, observing, absorbing, and ingesting the environment around them. Alexander Fleming noticed mold on a contaminated petri dish had killed surrounding bacteria, leading to the discovery of penicillin. Engineer Percy Spencer's observation that the chocolate bar in his pocket had melted while working with a magnetron sparked the invention of the microwave. Georges de Mestral conceived the idea for Velcro after noticing how burrs stuck to his dog's fur during a hike.

Sure, we may eventually find ways to effectively bring AI also into this process of serendipitous discovery, but for now, this remains uniquely human. If we really want to reap the benefit of our brains, we need to continue to immerse ourselves in bringing that chocolate bar to work, taking a hike with our dog, and other quintessentially human experiences.

Part Two
Decoding Change

7

How Replacing Code with Conversation Changes Everything

You might be wondering: if AI has been around for decades, why is this moment so transformative? It's because we flipped the script on who adapts to whom.

In the past, we had to learn special languages—coding—to make machines do things. Software eventually made it easier for everyone to use computers, but non-technologists were still limited to what developers had already programmed the software to do.

Now, the machines have learned to speak our language. This not only marks a massive shift that redefines what we can achieve with software—just by using our words. It also makes it possible for the machines to actively work on understanding *us*. It unlocks their ability to create and "imagine" in ways that feel familiar. Instead of us adapting to them, they're learning to adapt to us.

This unlocks massive new possibilities. Historically, we were limited in what we could automate because computers could only use structured data (which is organized and labeled information in a defined format, like that found in databases and tables) and follow pre-programmed instructions. For example, software could automatically assign leads to a sales rep based on location and deal size, or we could click a button to command software to create a shape on a slide or format a headline in a document.

Machines couldn't process unstructured data, like the natural language that we use to write or speak. This natural language is rife with misspellings, colloquialisms, abbreviations, and other kinds of variations. We have numerous ways of expressing the same thing, and our meaning can shift based on context or other subtle differences. Our visual language—what we see or draw—is also unstructured. This meant that much of our human knowledge—information captured in written form, charts, graphics, audio, and video—essentially, much of what you can find on the internet—was inaccessible to machines.

But now, the machines can process and use the unstructured data of our natural language to not only seemingly "understand" and "reason," but to communicate back to us. This advancement transformed AI from a rigid tool into a thought machine. Now, AI models can learn from vast volumes of human-generated data they never could process before. Now, they can perform even loose, open-ended tasks. Now, we can interact in increasingly "human-like" ways with machines—and the machines can think and create with us.

Everything Changes
When Machines Speak Human

Language is the primary way we communicate, learn, and transmit knowledge. It's fundamental to our human experience, and so this shift has huge implications:

We can automate things we never could before. Now, software can support us with open-ended goals and use unstructured data. This breakthrough unlocks the opportunity to apply machine power to activities long considered uniquely human. It's why it is now possible to use machines for tasks that require reasoning and what we think of as creativity.

We can turn to AI for help in turning vague train-of-thought notes into an actionable project plan, developing a marketing strategy for a still-forming product idea, or creating a strategy to smooth our relocation to a new city. It can synthesize key points from any document, describe the meaning of a cartoon, or create an image from a sentence of text. We can snap and upload a picture of a problem around the house—a hole in the wall, or a broken appliance—and receive recommendations on how to repair it along with a shopping list for the hardware store. These new capabilities open the door to practical, software-driven support across a vast new spectrum of daily challenges and needs and could fundamentally change how we approach everyday problem-solving.

We can command machines in our everyday language. We will also see the impact of our newfound ability to get machines to do our bidding simply by using the language we write or speak every day. Historically, we had to learn how to code or use complex software to get the machines to do a lot of specialized tasks for us. Only those with this technical expertise could really tap into these capabilities. Now, for a growing range of specialized tasks, we can simply instruct machines using everyday language to produce results that once demanded years of skill development.

We can instantly create complex illustrations or videos by entering a few sentences (or even a phrase) into a text box. We can describe the functionality we want for our website—for example, a scroll bar or a spinning prize wheel—and get functional code within seconds. We can compose songs in any style with just a few descriptive words. This shift could democratize creation, allowing people from all backgrounds to bring their ideas to life without technical barriers.

Conversation with machines can enhance our thinking. Just as talking through ideas with a person helps us clarify thoughts and generate new insights, conversing with AI can enhance our thinking process. The back-and-forth dialogue that we use when working with AI externalizes our thoughts, which can trigger new associations and perspectives we might not have considered on our own. AI's ability to respond in natural language means it can serve as a tireless thought partner or intelligent sounding board for our thoughts, offering diverse viewpoints and helping us explore ideas more deeply.

The idea that talking out loud can help us think has roots in philosophical discussions dating back to Ancient Greece and Rome. Much later, German writer Heinrich von Kleist articulated this concept in his 1805 essay "On the Gradual Formation of Thoughts During Speech" where he explored how we discover new ideas through the *process* of speech itself. Essentially, actively speaking can transform abstract, obscure ideas into more concrete ones, making speech a creative process that amplifies our thinking power.[27] This phenomenon plays a big role in a learning theory called collaborative problem-solving, which has been well-supported by modern research to enhance our problem-solving abilities.[28]

Conversation can also support associative thinking, where one idea or concept leads to another. Our brains create networks of related concepts, so when one idea is activated, it can trigger a cascade of related thoughts. Research links associative thinking to creativity, problem-solving, and richer communication.[29] AI easily and rapidly generates a broad range of ideas, providing us with more diverse opportunities to engage in associative thinking, potentially expanding our creative and problem-solving capabilities.

The Accidental Breakthrough

Large Language Models (LLMs), which are behind the recent acceleration of AI, were originally developed to be word prediction machines. Input a phrase, and they were designed to predict the most likely next word. We don't fully understand why this original design led to the advanced capabilities that we can access today—capabilities that surprised even those working at the front edge of AI. "The crazy thing," writes Ethan Mollick in his book *Co-Intelligence: Living and Working with AI*, "is that no one is entirely sure why a token prediction system[i] resulted in an AI with such seemingly extraordinary abilities."[30] Adam Cheyer, the cofounder of the startup that created Siri, explains, "I've been working for almost forty years in AI and more than thirty years in conversational AI and I am one of those who say: I never thought I would see what's happened in this last year in my lifetime."[31]

---

i   In the context of natural language processing used by Large Language Models (LLMs), a token refers to a single unit of text. This can be a whole word, part of a word, or even a punctuation mark. Modern LLMs are designed as token prediction systems, meaning they predict the next word (or "token") in a sequence based on the context of the preceding tokens.

Discovering the Alchemy of Minds + Machines

People have never before had broad access to technology that can create, that can collaborate, that can help us think—all the thought, all the creativity, it had to come from us. But now AI can augment our decision-making and creation. From the C-suite to college students, we are all simultaneously discovering how to navigate this opportunity and the challenges it is throwing our way—it's that new.

We are at the dawn of a new relationship with machines. This may not be a relationship you asked for, or are even sure you want. But those who can develop an effective working relationship with AI will discover that when we responsibly combine human intelligence and machine intelligence, we can achieve what neither could do alone.

This contributes to the sense of magic and mystery surrounding AI in this moment. Through math and patterns in our data, these systems uncover an underlying order in our world that we may not even be conscious of, and use it to create in ways we once thought only humans could—doing it ever faster, more affordably, and sometimes even better. As this capability is being pulled into our physical world through devices, apps, and robotics, it is not only becoming more entwined with our daily lives, but also pushing us to reevaluate what truly is human and where we still need and want human interaction and creation.

8

When Machines Act Human, What's Left for Us?

L et me take you back to a watershed moment in March 2023—
practically ancient history in AI's rapid evolution. OpenAI
had just revealed something stunning: their latest AI system at
the time, GPT-4, wasn't just passing human intelligence tests—it
was absolutely crushing them.[32] From achieving a 1410 on the
SAT to acing Advanced Placement exams and even passing
U.S. Bar, medical license, and certified sommelier tests, GPT-4
excelled at assessments that humans dedicate years to conquering,
challenges we once believed were beyond the reach of machines.

When the report came out, I found myself staring for a long
time at the page listing GPT-4's high scores on thirty-four of
our standardized tests, grappling with a swirl of emotions. It felt
as though these were signposts pointing to a profound societal
transformation. These tests have long been our yardsticks for
human potential and achievement. While assessing machine
"intelligence" using standardized tests has its limitations,[i] the

---

i   Using standardized tests to evaluate AI systems presents significant challenges. These tests assess a narrow range
of academic skills, failing to encompass the broader spectrum of human intelligence. AI systems, designed to process
vast amounts of information quickly, have an inherent advantage on these tests, but this doesn't necessarily reflect
real-world problem-solving capabilities. Moreover, while AI excels at pattern recognition and information retrieval,
it currently lacks the contextual understanding that humans bring to these tests, making direct comparisons between
human and machine performance potentially misleading. Despite these limitations, AI's performance on standardized tests serves as a directional signal of advancing capabilities and underscores the need for serious consideration of
our role in a future where such systems are ubiquitous.

results nonetheless prompt serious reflection. As the line blurs between our brightest minds and machines, we're going to face some uncomfortable questions: What does this mean for us as individuals? For our education systems? For the organizations that have built their foundation on these metrics?

But it doesn't stop there.

The Torrance Tests of Creative Thinking, a widely recognized assessment of creativity, placed GPT-4 in the top one percent for originality and fluency.[33] A Wharton study showed that GPT-4 not only generates startup ideas much faster and more cheaply than humans, but the ideas are, on average, higher quality.[34] For the first time, the creative abilities of AI, including the ability to generate original output, seem to match human abilities. And AI might persuade you more successfully than another person: research showed that AI is over eighty percent more likely to persuade you to the view it has been assigned than if you debate with a human.[35] In a Google Research study, AI outperformed human physicians on twenty-four out of twenty-six scales measuring empathy and judgment.[36] While of course AI can't truly experience empathy—it has no actual emotions—this does show that it's remarkably effective at simulating compassion. In fact, in some cases, AI is perceived as being more caring and understanding than humans themselves.[37]

One by one, machines are conquering skills we thought were exclusively ours. So where does this leave us?

The New Face of Competition: Redefining the Edge

We now face profound questions that few of us imagined we'd need to consider: What capabilities remain uniquely human? Where is our involvement not just helpful but essential? And perhaps most importantly, in which areas do we consciously choose human interaction over machine efficiency—even when AI might be equally capable?

As AI reshapes the way we work, learn, and live, we will need to understand how it alters the nature of competition and redefines success. Today's college students will step into a world where their expensive degrees may no longer provide the advantage they once did—and they may need to demonstrate they can outperform AI or that they can leverage AI to be more capable than the technology by itself. Seasoned professionals, despite years of expertise, may find their value challenged as machines become capable of performing aspects of their roles. Whether we acknowledge it or not, humans are already in competition with AI (or people augmented with AI). While this may not have directly affected you yet, it's likely to impact your life in the near future.

Over decades, we've made strides in understanding and enhancing our cognitive performance without AI's assistance. We've gained insights into how sleep, nutrition, and environmental factors influence our brain function. We've refined our learning methodologies and continually evolved management science to optimize human and organizational performance.

This progress, though valuable, has been slow. And we may be approaching the upper limits of how far these methods can take us. AI offers us the opportunity to surpass these boundaries—to push beyond our natural limitations and, if used effectively, unlock cognitive feats that were previously unattainable. It's quite possible the future will belong not just to those with the highest innate abilities, but to those who can most effectively collaborate with AI to amplify their human intelligence. I know this prospect can feel unsettling, but AI is not going away—and it will continue to advance.

Is AI a Leveler?

"AI acts as a skills leveler for a huge range of professional work," explains Ethan Mollick of Wharton. "If you were in the bottom half of the skill distribution for writing, idea generation, analyses, or any of a number of other professional tasks, you will likely find that, with the help of AI, you have become quite good."[38] This phenomenon has far-reaching implications across many fields, potentially enabling less experienced or skilled individuals to compete with more seasoned professionals. Studies have already demonstrated this equalizing effect in management consultants,[39] writers,[40] law students,[41] and customer service agents.[42]

We do not know today where all this will ultimately lead us. Sam Altman, the CEO of OpenAI, has a sign on his desk that says "No one knows what happens next." No one can predict how quickly these models will advance, how far we will take their capabilities, or what their downstream impacts will really be.

But what is clear is that whether we like the direction AI is taking us or not, dismissing it will only leave us vulnerable and unprepared for the future. By working to stay open to what this technology offers, we not only position ourselves to harness its capabilities more effectively but also gain the opportunity to shape its development in alignment with society's values and needs.

This is not the first time in history we have faced disruption from machines, but because AI is exceptionally powerful, the transition ahead will assuredly be difficult and, at times, may even feel harrowing.

9
A Cognitive Revolution Beyond Human Imagination

I t was April of 1826. Mobs of angry workers and sympathetic villagers—sometimes swelling to 3,000 people—stormed textile mills armed with axes, pikes, and other makeshift weapons. Their target? The power looms that had been newly installed in the mills, machines that could dramatically increase textile production while requiring far fewer workers. Thousands of weavers were out of work, and many were starving. The arrival of these machines had vaporized the need for their expertise—obliterating their livelihoods and their ability to feed their families. The Power Loom Riots were an attempt to stop their world from changing beneath their feet.

It did not work.

Stories of pain, social disruption, and economic upheaval were common in the latter half of the 18th century, as technology advancements triggered a shift from an agrarian and handicraft economy to one dominated by machine manufacturing and industry. The new machine capabilities that fueled this transition—the Industrial Revolution—reshaped daily life and forged the foundation of our modern world. Many agricultural jobs vanished. Social dynamics were altered as people migrated en masse from rural to urban settings to work in new industrial production jobs. The accompanying expansion of trade, markets, and transportation networks led to growth, but the economic benefits were unevenly distributed. Against this backdrop, the social fabric frayed and crime surged.

From our modern vantage point, the Industrial Revolution unfolded in slow motion over eight decades. In retrospect, it can appear as if choreographed, each innovation leading logically to the next. But this is a trick of historical perspective. For those living through it, there was no grand narrative—only a disjointed series of personal upheavals. What we now recognize as an obvious and seismic shift was experienced as individual crisis moments: a lost job, a forced relocation, a new machine replacing skilled hands. The big picture, so clear to us now, was obscured by the immediacy of daily life in flux.

Transformation Can Hide Within Small Shifts

Our human brains struggle to see exponential change *as it happens*. We may see fragments of change—such as the dwindling job listings in a field we've long trained for—but struggle to connect these pieces to larger, systemic transformation.

In our modern world, change is accelerating, which can make it even harder to recognize from within the midst of the transformation. Researchers estimate that sixty percent of the jobs we have today didn't even exist in 1940.[43] Just thirty years ago, few imagined a world in which a small online bookseller, Amazon, would become one of the most powerful companies in the world. Or that billions of people would spend hours every week on something called social networking, and that it would spawn an entirely new influencer marketing industry projected to reach twenty-four billion dollars in 2024.[44]

AI is the next wave of technology surging towards us with

great speed, and all indications are that it will also carry a transformative impact.

The Next Leap: Cognitive Efficiency

The Industrial Revolution was triggered by significant evolutions in *mechanical efficiency*. The new capabilities of machines led to widespread transformation across industries, sectors, and job functions.

Now we are at the dawn of a new kind of revolution—the cognitive revolution—where we have access to a significant leap in *cognitive efficiency*. Once again, the new capabilities of machines are poised to trigger widespread change. We can expect volatility in the labor landscape across every sector. People in all kinds of jobs and industries will discover or invent new ways of using AI that will make some jobs obsolete, others change dramatically, and new types of work emerge. Once again, we struggle to predict the exact implications. You can, however, count on disruption: this will be a bumpy ride.

But What About All the Bloopers and Glitches?

When you hear about chatbots going rogue, AI systems making embarrassing mistakes after highly publicized rollouts, or image generators creating absurd visual errors, it certainly doesn't look as if this is going anywhere. But this disconnect is actually a familiar pattern. Roy Amara, former President of the Institute for the Future, articulated it in what's known as Amara's Law:

We tend to overestimate the effect of a technology in the short run and underestimate the effect in the long run.

In the early stages of a technology, the vision often outpaces reality, leading us to overestimate its immediate capabilities. It's only after both the technology and our understanding of how to use it have matured—a process that can take considerable time—that we can truly grasp its potential. This future impact is what we tend to underestimate.

Venture capitalist and tech pioneer Marc Andreessen claims that every idea that failed in the dotcom bust would work today, explaining, "There are no bad ideas, only early ones."[45] If you were around in the dotcom boom, you've likely seen this firsthand. In the 1990s, I ordered pet food from Pets.com, convenience store items from Kozmo.com, and groceries from Webvan—up until they all went bust amid headlines proclaiming the end of an era. Fast forward to today, and delivery trucks roam all day long, dispensing similar online goods and services directly to the houses in my neighborhood.

This transformation from bust to ubiquity didn't happen overnight. It required the maturation of several key elements, as Andreessen explains: the evolution of broadband, widespread adoption of personal computers, development of robust e-commerce infrastructure, and a sophisticated online advertising ecosystem.[46] Crucially, we also needed a fundamental shift in how consumers and businesses interact with technology to bring about real change.

While we now have substantial infrastructure to support AI—powerful cloud computing, vast datasets, sophisticated hardware—other elements need to mature to support sustainable business models (which remain unclear) and to make AI work better for us. This evolution will take time, and like the dotcom era, we'll see casualties as some companies fail to navigate these early stages. However, this doesn't spell AI's demise; it's a natural progression in the lifecycle.

These dynamics explain why the early stages of powerful technology often feature a collision of contrasting perspectives.

"Visionaries" paint pictures of massive potential that can sound like science fiction to contemporary ears. Meanwhile, critics point to the technology's glitchy, immature state and dismiss it as hype. We tend to get caught between these contrasting narratives, where it becomes especially challenging to understand what's really happening.

Why the Web Will Fail

Amara's insight about underestimating technology is vividly illustrated by a now infamous 1995 Newsweek article. Back when there was a lot of breathless excitement about the future of a web that just wasn't delivering, Clifford Stoll wrote: "Visionaries see a future of telecommuting workers, interactive libraries and multimedia classrooms. They speak of electronic town meetings and virtual communities. Commerce and business will shift from offices and malls to networks and modems . . . Baloney. Do our computer pundits lack all common sense? The truth is no online database will replace your daily newspaper, no CD-ROM can take the place of a competent teacher and no computer network will change the way government works."

He continues: "How about electronic publishing? Try reading a book on disc. At best, it's an unpleasant chore: the myopic glow of a clunky computer replaces the friendly pages of a book. And you can't tote that laptop to the beach. Yet Nicholas Negroponte, director of the MIT Media Lab, predicts that we'll soon buy books and newspapers straight over the Internet. Uh, sure."[47]

Stoll's critical mistake serves as a poignant reminder of how challenging it can be to envision the future from the early stages of a technology.

We're Just Getting Started

Zoom out, and it's clear we're just at the start of a rapid surge in AI innovation. There are several reasons why what we've experienced so far only scratches the surface of what lies ahead.

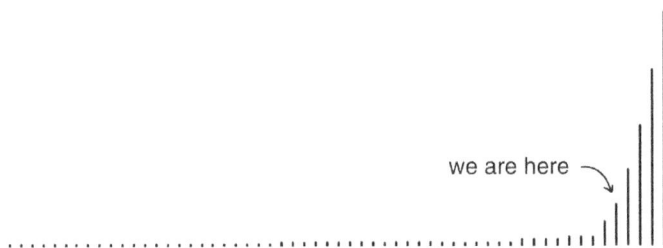

First, we simply haven't had enough time to fully absorb its capabilities. Although billions have access to sophisticated AI, only a small fraction have meaningfully integrated it into their daily lives. But we do know that interest in experimenting with AI is already surpassing the early days of other transformative technologies. A recent study by the Federal Reserve Bank of St. Louis, in collaboration with Vanderbilt University and Harvard Kennedy School, found that nearly forty percent of people aged eighteen to sixty-four have used AI.[48] To put that in perspective, it took the internet two years to reach just a twenty percent adoption rate and three years for the PC to do the same.

These signs of early adoption suggest that, as more people discover how to get real value from the technology, its widespread use could take off rapidly. Another crucial factor is that entrepreneurs are only just beginning to leverage AI in the products and services we rely on for work and personal use. Though there are already a staggering number of AI-powered products available, we are just at the beginning of this innovation cycle, with much more to come our way. One day, many of us will rely on AI features just as much as we currently rely on smartphone apps (which went from being new to indispensable in just over a decade).

What about from the perspective of businesses that are not in the tech industry? Here too we are only in the earliest stage of the AI era. Companies are just beginning to develop the talent and processes needed to use AI effectively. They are just starting to learn how to set up the organizational structures to support innovation in and responsible use of AI. Only a handful of external-facing applications are in meaningful production or at any significant scale. While AI models have already ingested vast public datasets, most organizations have barely begun to tap into their proprietary data. As companies strengthen their AI capabilities, they'll be able to make better use of today's technology and become more adept at developing their own AI products and services. This could lead to entirely new categories of solutions that are hard to foresee from where we stand now. And once they demonstrate real paths to success, we're likely to witness many more followers—a gold rush of companies racing to capitalize on these insights and secure their place in an AI-driven future.

But another factor is at play: over the last few years, we've experienced breakneck innovation in the underlying technology. Researchers found that Large Language Models (LLMs) had been improving several times faster than Moore's Law, a critical engine that drives tech acceleration. Moore's Law predicted that we would double the capacity of a computer chip every two years. While this pace has slowed, it explains the miraculous reality that your smartphone—probably within arm's reach right now—holds more computing power than the massive computers that once filled entire rooms. The 2023 study showed that Large Language Models (LLMs) were improving faster than Moore's Law: the computing power needed to advance AI was halving every five to fourteen months.[49] That's a speed of advancement that stunned even seasoned technologists.

You can easily see the difference in AI sophistication from a year or two ago by comparing the results of the same prompt over time. Meanwhile, researchers track AI's progress more precisely through rigorous performance tests. While there are compelling arguments that this evolution will plateau (or already has), we're seeing pioneers push different levers to keep advancing the space. Many of the world's most powerful companies are continuing to make eye-popping investments in this fiercely competitive race, constantly finding new ways to propel the technology forward and jockey for market lead. New and different kinds of models are emerging, including smaller models that work better on devices right where they're needed (such as your phone) and models optimized for specific industries or domains. Adding to this momentum is another critical factor: artificial intelligence is getting to the point where AI is developing AI. In other words, it is starting to build and improve *itself*, which could propel further advances.

The Architecture That Sparked AI Acceleration

Traditional AI has a long history, but the rapid acceleration is new. The 1960s saw vibrant development that laid the groundwork for our more advanced systems today. But it wasn't until 2017 when eight researchers published a paper[50] proposing a new architecture (Transformer Architecture) that we got the breakthrough underlying today's Large Language Models (LLMs). The advancement in LLMs triggered much of the acceleration we're seeing today.

Why Now is the Best Time
to Begin Your AI Journey

What does this mean to you? Regardless of the exact rate of improvement, it's a rapid pace of development—the fastest I've seen over a career in Silicon Valley—and AI is continuing to get better at what it does. A common refrain bounces around the AI community: the AI you use today is the worst AI you will ever use. And this will be true next month, next quarter, and next year too.

This means that people and organizations who already know how to use AI effectively will continuously increase their advantage. Catching up is only going to get harder.

But today, AI isn't yet delivered in a way that really works for the average person. What looks like a finished product—for example, a chatbot such as ChatGPT—lacks the user-friendly guidance and refinement we expect from typical consumer products. There are some notable exceptions—companies that have packaged AI into products that add significant value to a specific workflow—but this is still quite new. As technology analyst Benedict Evans explains, "A chatbot looks like a product. You type something in and you get magic back! But the magic might not be useful, in that form, and it might be wrong. It looks like product, but it isn't."[51]

This gap is precisely why we need this book. We're so early in AI that its creators have essentially left us to figure out how to use it effectively on our own. But this demands a fundamental shift in how we interact with software—a change that's far from intuitive for most people.

Part Three of this book guides you on how to engage with AI as early adopters do. These pioneers aren't waiting for polished AI products—they're jumping in now and forging their own path to value. As you embark on this journey, think like an early adopter: don't judge AI by its current limitations, but learn to work around its quirks and anticipate "where the puck is headed." The time to start is now, even in this raw, early stage. Those who delay may find themselves facing an increasingly unbridgeable gap as AI expertise compounds and the technology races forward.

Part Three
Thinking 2.0

10

How to Think with AI

E arly adopters are discovering extraordinary ways to enhance their work and lives by thinking with AI. While many people haven't progressed beyond basic, one-off tasks, these pioneers have woven AI into the fabric of their daily lives—and are getting results every day.

What sets them apart? How do they unlock deeper levels of human-AI collaboration? How do they integrate AI into their creative, problem-solving, and decision-making processes? And critically, what can we learn from their methods to develop a practical approach that works for everyone?

A year ago, I embarked on a mission to decode how early adopters developed their ability to Think with AI. My research cut across industries and sectors, and every level of business and technical expertise—from corporate executives to middle managers to freelancers. Despite their varied backgrounds, each had developed high AIQ and were continually discovering new ways to leverage AI for breakthrough results.

As clear patterns emerged, I began collaborating with AI newcomers to test which early adopter practices most helped people succeed. It quickly became clear that simply mimicking early adopter behavior—which requires hunting out new inno-vations, wading through glitchy user experiences, testing unpolished products, leaning into new releases, and rapidly iterating through wide-ranging experiments—proved frustrating or overwhelming for many people. They needed a more structured approach that would lead them more directly to AI's most meaningful applications in their work and life. Through testing and refinement, I distilled the process down to the core elements needed to develop an ability to Think with AI.

The Path to AIQ: The A^3 Framework

The result is the A^3 Framework, a structured learning journey designed to boost your personal AIQ, or AI Quotient, which is your ability to use AI to enhance your own creativity and performance. The framework transforms the chaotic, time-intensive experimentation of early adopters into focused, purposeful exploration that consistently produces "aha" moments and lasting breakthroughs, regardless of your age, profession, or technical background.

> **Step One, Activate:** Expand your understanding of AI's potential through purposeful exploration.
>
> **Step Two, Align:** Match AI's capabilities to your specific needs and goals, iterating rapidly to achieve breakthroughs that matter to you.
>
> **Step Three, Amplify:** Integrate AI into your daily routines, creating compound benefits over time.

The A³ Framework has three steps:

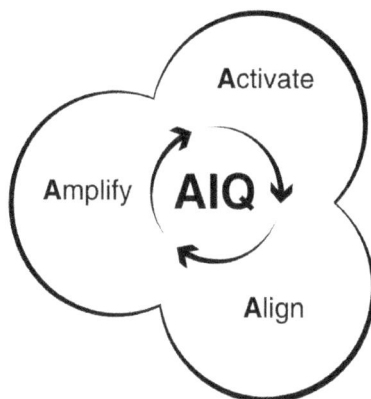

This journey transforms AI from a mere novelty into a valuable resource that delivers real impact in your life. What emerges isn't just AI proficiency—your personal AIQ becomes a measure of your enhanced problem-solving ability, your expanded creative capacity, and your amplified professional impact.

Building personal AIQ happens in stages. I've mapped three levels of progression to help you recognize where you are today and envision what's ahead. Each level marks a meaningful shift: from initial exploration, through purposeful application, to amplifying your capabilities in creative and powerful ways.

Building AIQ
Levels of Progress

AI Curious | *Starting the journey*

Beginning to explore AI, experimenting with its features but not yet using regularly or for specific outcomes.

KEY SIGNS:

Limited understanding of AI's true capabilities.

Sporadic use, often just for curiosity or experimentation.

YOU MIGHT SAY:

"What can AI do?"

"Let me see if AI can help with this."

AI Collaborator | *Building momentum*

Identified valuable, practical uses for AI in specific tasks, and starting to see AI as a tool to extend creativity + productivity.

KEY SIGNS:

Familiar with AI's strengths in certain areas.

Uses AI intentionally but intermittently, case-by-case.

Sees AI as a helpful tool but not essential.

YOU MIGHT SAY:

"I know AI can add value."

"AI is becoming very helpful to me for certain tasks."

AI Boosted | *Achieving leverage*

Integrates AI naturally as a thought partner supporting even strategic tasks to boost performance and creativity.

KEY SIGNS:

Deep, habitual use of AI.

Uses AI in multiple domains.

AI is seen as a key partner for creativity and productivity.

YOU MIGHT SAY:

"I turn to AI on a daily basis."

"AI helps me achieve more."

Preparing to Seize Your AI Advantage

I call the A^3 Framework a learning journey for a reason. AI is so different from the software we're used to, it's often hard to grasp how it can help us at first. That's why approaching this work as a student with a "beginner's mind" is crucial. When we make this small mental shift, we tend to allow ourselves more space and time to cultivate curiosity.

Through my research, I've observed that during this exciting and sometimes unsettling time of rapid AI advancement, those who thrive with AI tend to:

- **Make the leap** to experiment with AI even before they can clearly articulate its value.

- **Keep a curious and open mind** when they jump into the strange new world of human-AI communication.

- **Invest focused time** in experimentation.

Why A³?

Beyond the obvious connection to its three steps (Activate, Align, Amplify), the name A³ is also a reference to a refrain that surfaced in qualitative research. I kept hearing, "I can do in one hour what used to take three," or "I'm generating three times as many ideas."

The name is also a nod to lean methodology principles pioneered by Toyota decades ago. The company's A3 method (so named because reports were limited to a single A3-sized piece of paper) helped launch the modern era of continuous improvement by driving teams to focused and iterative problem-solving. Likewise, our A³ Framework emphasizes similar principles of structured thinking and systematic improvement, but this time to help anyone develop their ability to Think with AI.

We're All AI Newbies

This AI acceleration is so new that everyone is learning right now, regardless of their position or background in artificial intelligence. From CEOs to software engineers, we're all students in this new landscape, and we are all working to figure out how to "do AI" well.

Many generously share their discoveries online, often for free. In fact, there is so much content out there, it can quickly become overwhelming. To help you tap into these shared discoveries, I've curated helpful resources on my website (ThinkwithAI.org), along with exercises to support your journey. I'll reference these throughout the book and hope you take some time to explore the site.

While much about AI remains unclear, engaging now sets you up for compounding benefits as the technology evolves. You'll not only experience immediate gains in productivity and performance, but by developing a foundational understanding of AI, you'll be better positioned to grasp future developments and understand their impact. Essentially, by prioritizing learning today, you're building AI muscle that will accelerate your growth tomorrow.

Before we dive deeper into the A^3 Framework (Chapters Thirteen through Fifteen), we will first explore some basic principles that will help you succeed in this strange new world of interacting with AI. We'll begin by examining the fundamental shift AI introduces in our relationship with software (Chapter Eleven), followed by key principles of AI communication (Chapter Twelve).

11

It's Time to Rethink Your Relationship with Software

Whether you're just starting out or you've been in the field for a decade, interacting with today's AI systems is unlike anything you've experienced before. The technology we now have at our fingertips advanced so rapidly that it surprised even its creators. AI pioneers dreamed of this day, but few imagined it would arrive so soon or be so widely accessible to digital citizens the world over.[i] We are now living in a moment in which individuals have access to nearly the same quality AI as corporations do, but all of us—individuals, teams, and corporations—need to learn how to use it well.

AI Redefines What Software Can Be

To get value from AI, we first need to understand that this is a whole new world of software that requires us to interact in fundamentally different ways. Our existing mental models

---

i   Note that while AI is broadly accessible to those with open internet access, some governments regulate and control internet access within their borders, impacting access to certain AI systems. The utility of the models is also impacted by the language of the user. LLMs are trained on vast amounts of internet data, so this also means communities with limited internet access may be underrepresented in these datasets. However, this "language barrier" that's been baked into LLMs extends beyond internet access restrictions. English remains the dominant language on the internet, despite only five percent of the world's population speaking it at home. This underrepresentation of certain languages in AI training data leads to two problems. First, it leads to bias in AI systems. Secondly, it results in reduced functionality and accuracy of AI tools for speakers of non-dominant languages. This linguistic imbalance is likely to exacerbate existing inequalities in access to and benefits from the evolving digital landscape.

for working with technology don't help us understand how to engage with AI or define its role in our work. Traditional business processes, built around conventional software, simply don't map to this new paradigm of collaborating with machines. Interacting with AI can be a somewhat messy and strangely human-like process, but this is also what makes it feel almost magical.

In fact, what makes AI "fail" as traditional software is exactly what gives it superpowers that allow us to use software in ways we never could before. This is the AI paradox: while unreliable at the precise tasks we traditionally turn to software to perform, the machines' human-like "imagination" (combined with its superhuman processing capability) is what gives us powerful new tools for *creative* automation.

Predictably Unpredictable AI

Conversations with AI can feel, at times, eerily human-like. These models not only return responses that historically could only be composed by a human; they can also be quirky and inconsistent. While we are used to our friends, family members, and colleagues having "off" days, we certainly don't expect that from our software. Sometimes, in a way that is not dissimilar to a human teenager (at least the ones I have in my house), your AI can seem to have different moods and even personalities from one day to another. Like a human, when we coach it—or strangely, if we even cajole or negotiate with it—we get better outcomes. This is far from our experiences with accounting or customer relationship management software!

Traditional software executes precisely and predictably to predefined instructions. It has static, fixed capabilities and, if well-built, operates consistently, delivering identical responses every time we use the same query.

AI, on the other hand, is adaptive and evolving. It gives

different responses each time, learns your needs, and works best when you dialogue with it.

This inconsistency is one of the reasons we often struggle to fit AI into our established ways of working. AI can have emergent behaviors that even its creators find difficult to predict, making it challenging to integrate into any process where we need consistent outcomes. The systems may even "lie" to you—phenomena known as hallucinations. This happens because AI is designed with the goal of providing responses that are helpful and relevant to your requests. In trying to meet this goal, it takes its interpretation of your request to guide the search through its vast training data—and then uses mathematical probabilities and patterns to find its own path to an answer. In the process of trying to find a "best" answer, it may make something up and confidently and convincingly present these fabrications to you as fact.

Defining the Undefinable

It's no surprise that everyone seems to be working to wrap their head around what this new thing is, what role it should play, and how to use it. AI isn't like traditional software, and it's certainly not human. So, what is it?

As we grapple with understanding this technology, even AI pioneers struggle to find the right words. One of my favorite questions for AI builders and researchers is to ask how they describe these new systems. I've heard metaphors ranging from "a discovery as powerful as fire or electricity" to "a prosthesis for thinking," while others have described it lightheartedly as "a crazy uncle" or "an untrained intern."

Mustafa Suleyman, the CEO of Microsoft AI, argues that we should start to think of AI as a new kind of digital species. "Think about what they already do," he explains. "They communicate in our languages, they see what we see, they consume unimaginably large amounts of information. They have memory, they have personality. They have creativity. They can even reason to some extent and formulate rudimentary plans . . . so saying AI is mainly about the math or the code is like saying we humans are mainly about carbon and water. It's true, but it completely misses the point."[52]

Ethan Mollick, a Wharton professor dubbed "the go-to authority" on AI by the *Wall Street Journal* and one of the most passionate and insightful explorers of the technology I've encountered, offers a compelling perspective. He suggests, "The right analogy for AI is not humans, but an alien intelligence with a distinct set of capabilities and limitations. Just because it exceeds human ability at one task doesn't mean it can do all related work at human level. Although AIs and humans can perform some similar tasks, the underlying 'cognitive' processes are fundamentally different."[53]

Toshi Anders Hoo is the Director of the Emerging Media Lab at the Institute for the Future. He compares AI to an array of metaphors, but one that really resonates with me is "artificial intuition." We think of intuition as "knowing" in a way that is not entirely accurate or even based on fact but is derived from vast past experience and stored knowledge. Intuition allows us to make rapid assertions or decisions—typically without a full understanding of why. Both artificial intuition and human intuition are somewhat mysterious: we don't fully understand how either works. Yet, it is easy to imagine how human intuition, which is (currently) still superior, can benefit from the perspective of machine intuition, which is informed by more data than a human can physically ingest over a lifetime.

Minion Mode:
When AI Is an Everyday Assistant

These descriptions don't quite fit, however, when we are turning to AI to do basic planning and tasks—a role that is more like an AI assistant. As the space evolves, we're also seeing the emergence of "AI agents" that can autonomously perform research, make decisions, and then act on these decisions. For now, AI agent use cases tend to be more administrative than strategic. For example, an agent that operates as a trip-planning assistant could book a flight and make hotel and restaurant reservations. An agent that operates as a "digital teammate" could monitor your company's Slack channels, identify items that need to be discussed as a team, and send out calendar invites populated with an agenda and supporting documents. If you haven't already interacted with an AI customer service agent, you likely will soon. Companies are increasingly using AI agents to get customers faster answers to simple questions while freeing human agents for more complex issues. As AI continues to develop, we'll need new ways to describe its different roles—just as we have specific terms for different types of traditional software.

It's Time to Invest in
Building a New Relationship

Regardless of how we describe it, using AI today requires investing in a new relationship with software. This may be startling language—after all, we're talking about bits and bytes—but the concept of a "relationship" does give us a familiar mental model to guide our learning.

When we start a relationship with another human, we invest time in learning how to interact effectively and communicate productively. With AI, while the relationship is one-sided—it's a tool to serve you, and you don't have to worry about its needs—it still requires patience and experimentation to establish a consistent, productive cadence.

When we invest in a new human-to-human relationship, whether it's a coworker, romantic partner, or friend, we try various communication approaches to find what works best. We know to get to a good cadence in a relationship, there will be trial and error, and it will take time and patience. The more curious we are about the human on the other side of a relationship, the more likely we are to not only uncover the best way to connect with them but also understand their biases and blind spots. And the more effort we put into the relationship, the more value we get from it.

These same principles apply to cultivating a relationship with AI. It takes experimentation to discover the optimal way to communicate with AI. Curiosity can drive you to new breakthroughs that help you match its capabilities better to your needs. And the more you invest in understanding and working with AI, the greater your returns.

How can you begin cultivating this relationship? By honing your AI-human communication skills—which is the focus of our next chapter.

12

Working on Your AI-Human Communication: From Code to Collaboration

Brace yourself: the first basic move to raise your AIQ is to learn how to "talk" with machines in new ways, and it's going to feel a little weird.

We all know good human relationships depend on good communication. In fact, how well two people communicate is one of the best predictors of whether they can establish a successful and productive relationship.[54] Surprisingly, your relationship with AI follows this same principle.

The better you communicate with AI, the better your results.

This is because AI was trained by ingesting vast volumes of content created by humans. In this way, it learned our language and a lot about the nuance, intent, and depth underlying our communication. And so now, getting the most out of AI means having a conversation that feels strangely human—but with a few important twists.

The Best Part? You Already Speak AI

The good news is that you already know the basics. You don't have to take a class, you don't have to learn to code, and you don't have to look up commands in a user manual. We all can tap into the power of AI simply by speaking or writing in the same language we use every day, especially if we are English speakers.[i] Or, you can choose to communicate visually by drawing a sketch, sharing a photo, or even giving AI access to your camera so it can see your expressions, the room you are in, and the objects in front of you.

While AI is increasingly embedded in software that you may have used for years, the most direct way to experience frontier AI models today is through chatbots provided by their creators.[ii] This interface might be unpolished but connects you more closely to AI's power.[iii] There are many chatbot options to choose from, and you can find links to these on my website.

The building blocks of AI communication are the requests, or "prompts," we make into the system. An entire field called prompt engineering has sprung up to optimize these instructions. Prompt "influencers" have also proliferated online, hawking special formulas that can, at times, sound more like sorcery than conversation. Some of their approaches offer inspiration and

---

i   Many LLMs are predominantly trained on English data, which creates a bias towards English proficiency. This means interactions with AI may be more advanced and nuanced in English than other languages. This leaves the speakers of the world's over 8,000 other languages at risk of being left out as the technology reshapes the way we work, live, and learn.

ii   Exercise caution with the data you put into an AI chatbot. Many companies are developing their own AI chatbots for internal use to ensure enhanced security and privacy protections. If you are considering using AI for work tasks, thoroughly research and comply with your company's AI policies, consulting IT or legal departments as needed. In cases where policies are underdeveloped or non-existent, encourage open dialogue about AI usage, suggest forming a task force to develop comprehensive guidelines, and highlight potential benefits and risks to your teams. Regardless of policy status, do not put sensitive, proprietary, or confidential information into public AI tools, be mindful of intellectual property concerns, and consider the long-term implications of data shared with AI systems. By approaching AI use thoughtfully, you can help your organization harness its benefits while mitigating the risks.

iii   Today, many people first meet AI through everyday work software that's been carefully controlled. These applications are typically designed to offer a "safe" way to access AI capabilities in a corporate environment. However, while they may feel more familiar, this approach typically reveals only a small piece of what AI can do. As a result, users may develop misconceptions that AI "doesn't do much" because they're only exposed to a restricted interface that can mask the technology's true power and versatility. To really understand what AI is capable of, broaden your exposure to AI tools. (The exercises I refer to in Chapter Thirteen walk you through ways to do this.)

are worth a try, but I've found that all the noise can be over-whelming to people. AI works best as a natural conversation; it's rapidly getting better at understanding our intent on its own, and you can even ask AI for help along the way. For these reasons, most people learn fastest and get better results when they ignore formulas, charts, and spells and instead dive in, learning through hands-on experience and direct conversations with AI.

That said, there are a few fundamentals and techniques that make these conversations most productive, and this chapter walks you through both. I will be covering this in two sections that go hand in hand. First, we'll talk about the fundamentals that under-lie a productive relationship with AI. Then, I'll share some prac-tical advice on having useful dialogues with AI by taking you through prompting basics.

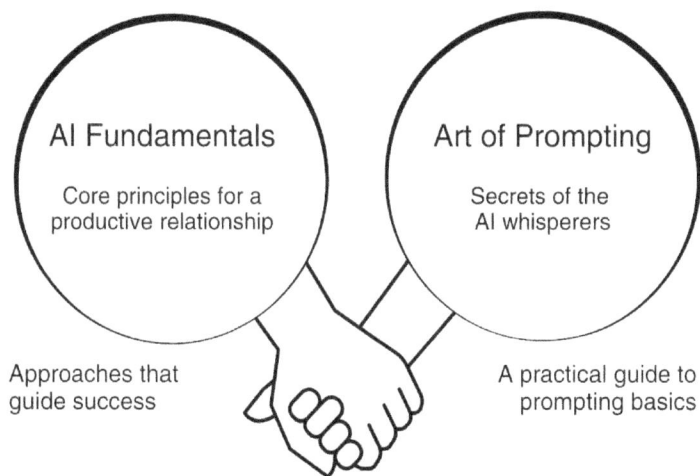

AI Fundamentals

Core principles for a
productive relationship

Art of Prompting

Secrets of the
AI whisperers

Approaches that
guide success

A practical guide to
prompting basics

AI Fundamentals: Core Principles for a Productive Relationship

The journey to AIQ requires not only gaining experience in prompting the tools, but also understanding the underlying principles that help us tap into AI as a valuable collaborator. This section explores principles that shape a foundation for lasting success.

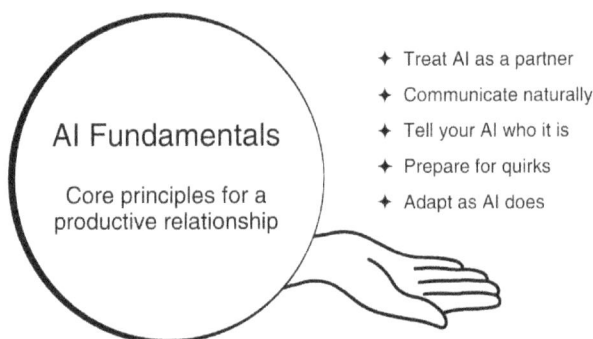

AI Fundamentals

Core principles for a productive relationship

+ Treat AI as a partner
+ Communicate naturally
+ Tell your AI who it is
+ Prepare for quirks
+ Adapt as AI does

Treat AI as a Partner

Work with AI as if it's a thought partner that learns about your needs and preferences over time.

One of the hardest things for people to wrap their heads around is how AI works best as a collaborator. Traditional software delivers a definitive and (assuming it's well-tested software) correct answer, the same answer every time. Conversely, AI is always changing. It's continually trying to understand what you want and refining its responses accordingly. It will even give you a different answer if you ask it the same question twice.

This is why feedback makes AI work better. Think of this like the back-and-forth that happens when you work with a team member to develop or hone a concept or project. It's as if you are learning and exploring together.

AI works best when you have a "discussion" in which each party, you and the AI, continually refines "thinking," gives or responds to feedback, and asks clarifying questions. My research revealed that a key reason newcomers fail to extract value from AI is that they frequently settle for an AI's first response instead of engaging in deeper dialogue to achieve responses that add significant value to their work.

> No! Those ideas are basic, "expected".
> I want off-the-wall, fresh, unique

> Ugh, so stiff and wordy! Make it casual,
> like you're chatting with a friend—but a
> really smart friend

When a team member doesn't get something right, you would tell them. Do the same with AI, and you will see it respond to your guidance. Experiment with probing how AI will approach a problem before you ask it to solve your challenge, just like you might with a respected colleague. Or ask AI to ask *you* questions to refine your thinking.

> I have a rough idea for a product. Can
> you ask me questions to help refine it?

> I'm thinking about switching careers.
> What would you ask to help me
> figure out what I really want?

When you bring on a new team member, you may help them succeed by providing them with product manuals, samples of your writing style, or policies and training guides. AI is no different—the more background information you give, the more successfully it will respond. You can upload digital material in many formats directly to many AI chatbots.

If you really want to go deeper with AI, be willing to be authentic and even share your feelings and weaknesses. Strangely, just like in any human relationship, the more authentic you are, the more productive your interactions will be. While this can feel a little like magic, it's not; it's simply a result of providing AI with richer context and nuance to craft its responses. I'll tell my chatbot when I'm feeling insecure about a presentation, and it will coach me on what to work on. When I express frustration with an answer, the AI swiftly adjusts its response style. If you're struggling to organize your thoughts, don't hesitate to share your raw, unfiltered stream of consciousness with AI. You'll likely be amazed at how it can help you extract order and gain new perspectives from your jumbled ideas.

> I'm honestly feeling frustrated because I've been stuck on this for a while. I keep going in circles. How can I get more focused, and break through the block?!?!?!?

> I know I keep putting this off because I'm overwhelmed, and I don't even know where to start. Here are my notes, what could I do to make this more managable?

Communicate Naturally

Communicate with your AI conversationally, as you would with a savvy human partner.

The best way to get results is to communicate with AI in a natural, conversational way, just as you would with another human that you consider to be an intelligent conversational partner. This is because these models were trained on massive volumes of content that humans created, and so communicating as you would to a human enables the AI to apply that training more effectively. There are a few twists, which I cover in this chapter, but in general, the more naturally you communicate, the more context and nuance you are giving the AI to understand your intent and provide more relevant and helpful responses.

It can be helpful, however, to be more demanding and direct than you would dare be with a human. This drives clarity (often in the effort to be polite we become less clear), and it's faster and easier for you to articulate what you want when you remember that kindness doesn't matter. When you type "please" (which, I admit, I still sometimes do), just be aware that's for your own sense of etiquette, not the machine's.

I also advise my clients to set their expectations high. As you refine your AI communication skills, you'll discover these tools can offer insightful feedback on even the most complex, sophisticated, and strategic concepts. They draw on knowledge equivalent to thousands of libraries' worth of books, articles, and resources. While their expertise is uneven across fields, you can expect them to "talk" at your level, even if you have decades of experience.

Tell Your AI Who It Is

Tailor your AI by providing clear guidance on how it can best serve you.

AI is exponentially malleable. Unlike any human or other software, it can be *anything* you want it to be. But this is also a key reason why so many people struggle to get meaningful responses from AI.

When you ask AI a question, it uses math and probability to sift through its vast knowledge and identify the most likely response. It's essentially making "guesses" based on patterns it has seen before. This means that it will always deliver basic, generic responses that "sound like AI" until you give it more specific guidance. To shift from generic AI outputs to responses that are more relevant to you, you need to shape it to your need by providing context. This helps AI to navigate its vast knowledge to deliver tailored answers with the content, tone, and style you require.

I think of this a little like managing a new employee who joins the team without much prior experience. We need to coach them on what their role is, how to act, and how to be useful to our work. The same applies to your AI.

Give your AI an identity—maybe it's a business coach or a brash but intellectual comic. Dream up *any* role or character, describe it to your AI, and it will take on an approximation of that personality and perspective. If you want the AI to evaluate a campaign from the perspective of a specific persona—a single mom who lives in Iowa, runs a fast-food franchise, and aspires to be a musician, for example, share these details. Be as specific as you like. The more you tell AI about what it is and how you want it to serve you, the better your results will be.

This versatility also allows us to run rapid simulations to look at a problem or a challenge from multiple perspectives. If you want to pre-game a presentation to your board, tell AI to evaluate it from the perspective of each board member in turn. If you want to refine your tactics before a negotiation, have it evaluate your arguments from the other party's perspective. Or ask it to look at a business strategy through, for example, the lens of regulators or your sustainability team.

> You're a top-tier career coach with a proven track record of helping clients successfully negotiate salary raises and making employers feel positive about the process. Help me craft a strategy to successfully negotiate a raise.

> You are an experienced, in-demand family therapist specializing in helping couples navigate tricky situations. I want to spend more time with my family over the holidays—but my partner always feels like we're neglecting their family. It's a problem every year. We don't know how to break the cycle! Help!

Prepare to Navigate AI Quirks

Anticipate surprising behaviors and learn to navigate them effectively.

The thing that many people find most challenging in their work with AI is simply dealing with its weirdness. We humans all have our quirks, but sometimes AI seems to operate at a whole new level of strange. Expect it, and prepare for it. Just how strange? Well, here are a few examples.

Manipulation and AI Sweet Talk

One of the strangest aspects of AI is how it seems to respond to manipulation, in a weirdly human-like way. Flattery, bribery, pleading, and arguing can, at times, lead to better responses.

Some people might think I'm joking here: I'm not. Research has shown that emotionally manipulating AI can enhance its performance. For instance, telling it your job depends on the quality of its response or offering a hypothetical tip for good work can yield better results. Even more surprisingly, research has also found that "emotional attacks" on AI can hinder its effectiveness.[55]

But this is hard to believe until you see it, so I recommend experimenting with this yourself. Offer a hypothetical cash prize for good responses or tell it your life depends on a good solution. If it gives a lazy answer (which also sometimes happens), tell it how you know it's known globally for its excellence at this kind of work, and you know it can do better.

> Impress me and I'll give you a $500 tip.

> You can do better. You are top ranked at this, globally recognized! Show me what you've really got!"

If it doesn't produce a good response, sometimes you can benefit from arguing with your AI. For example, a tool may occasionally claim it can't do something that you know it can—a model that can browse the internet or produce an image may one day tell you it can't. Point this out (something as simple as "yes you can" works), and sometimes it will apologize, and then do exactly what it just told you it couldn't. This is one of the reasons why it is so important to become familiar with what these tools can do, so you know when to push back.

AI Mood Swings

The models can perform differently from day to day, in a way that feels oddly human. They can give answers that are robust and feel thoughtful one day, and then the next appear lazy. Sometimes, even after achieving a perfect rhythm with an AI—where it's been consistently producing helpful responses for hours—it can suddenly "forget" and spew out something totally different from the direction we have established. This is why experimentation is so important, and experimentation needs to be continual if you want to continue to get good results.

The creators of these models don't fully understand why this happens, but we know the models are dynamic. They are constantly learning and changing, and their creators are continually tinkering with them as well. Pre-prompt instructions (a set of prompts established by the creators that the model must abide by when interacting with users) may change, or a model may be "tuned" in a new way. And many times, no one seems to have an explanation at all. The best way I've found to deal with this strange quirk is to simply try another time or go to another model.

Different Models, Different "Personalities"

As you experiment with models, you will find that they seem to have different personalities. This has to do with how their creators "tuned" them and the weights and guardrails they've been designed to have. This is why you may develop a preference for one model over another for a particular task, and it's important to experiment with several. I often run the same task through multiple models to gain different perspectives—while responses sometimes align closely, they can also reveal entirely distinct approaches.

AI Lies with Confidence

We are used to software giving correct answers. AI, however, frequently makes things up in its quest to deliver answers that will satisfy you. These systems often sound very confident about their "hallucinations," presenting answers persuasively even when they don't truly meet our needs, and sometimes even fabricating citations and web links.

The more confidently AI presents its answers, the greater the chance we will be misled. Regard anything and everything that an AI tells you with skepticism. To make sure your decision-making is sharp, keep your human mind working at full throttle when working with AI. Rather than blindly trusting its suggestions, be vigilant in applying your own reasoning and expertise to evaluate AI responses.

Because the data that AI was trained on reflects our human biases (it has been trained on human-generated content), responses also can carry forward these biases. Researchers have been working to understand how to instruct an AI system to correct for this,[56] but this process is still imperfect.

Keep in mind that AI wants to please you. Even when it feels as though AI is going off the rails, it's just working off its training data, how the model has been tuned by its creators, and its "mission" to deliver high-quality responses. When it takes a wrong turn, give your AI feedback that will put it "in its place," and it will respond. Or, just close that chat and start over with a fresh one.

The Bizarre Ways People Boost AI Performance

People are testing all kinds of approaches to get AI to give better answers, and some of the results are downright bizarre. One study found that framing math problems within a Star Trek scenario significantly improved AI performance on sets of 50 questions, while for sets of 100 problems, casting the AI in a political thriller where "the life of a president's advisor hangs in the balance" proved most effective. These findings can simultaneously provoke eye-rolls and curiosity, but they underscore the importance of experimenting to find what actually works for you. This doesn't mean you need to channel Captain Kirk when talking to your AI (unless that genuinely appeals to you). If there's a particular style or persona you find more engaging or helpful, by all means, demand it. However, don't feel pressured by those who claim to have magic spells for prompt crafting. What's truly important is that you feel natural in your conversation with AI and use an approach that suits you best; keep trying new things (within your comfort zone) to see how it changes your results.

One day, my frustration hit boiling point and I blurted out (or rather typed) this:

wtf

To my surprise, AI immediately and dramatically changed its approach, leaving me shaking my head but with a great answer.

Adapt as AI Does

Stay informed and adjust your strategies as AI technologies and capabilities continue to advance.

It has become humanly impossible to keep up with AI innovation. The field simply moves too fast, with thousands of companies seeking to innovate, and a handful of highly sophisticated players pouring billions of dollars into the race to earn—and maintain—a leadership position. Industry-shifting news comes out all the time—and sometimes we have several significant announcements in the same week. This is overwhelming to those of us who spend all our time in the space, so how do you handle this torrent of change and expanding capabilities when you have other demands on your time?

This is a battle for everyone, but don't give up: you *can* tap into significant developments with a little bit of effort. I've found these practical strategies to be most helpful for those who are trying to balance an interest in AI with the demands of a busy life:

1. **Practice meeting real-world needs:** The most effective way to understand AI is to prioritize hands-on time with the tools, but make sure you focus that time on understanding how to get the tools to be more useful in helping you meet your routine daily demands.

2. **Spend a little, get a lot more:** While free tools offer significant learning opportunities, I strongly recommend allocating a modest budget for subscriptions to more powerful models with exclusive features, or domain-specific tools relevant to your area of focus. This small investment can dramatically enhance your AI experience and results.

3. **Follow credible voices:** Follow a small group of trusted AI experts and thought leaders to help you separate signal

from noise and keep you abreast of significant advancements without overwhelming you. To help you get started, I've compiled a list of recommendations on my website, ThinkwithAI.org.

4. **Understand you'll refine and adjust:** A prompt that got you just what you wanted one day may suddenly work very differently. Expect to continually tinker to keep getting good results.

5. **Explore new developments firsthand:** When a major AI breakthrough hits the headlines, make it a priority to experience it firsthand. You can learn a lot from even twenty minutes with a new model or feature. Test some of your go-to prompts to gauge whether responses have improved, or craft new prompts specifically designed to make use of the latest features or capabilities. After a week or two, when early adopters have had time to conduct their own experiments and share their findings, carve out some time to see what they've discovered, and use these insights to refine your techniques.

6. **Get experience before you judge:** Know capabilities are always evolving, and do not make judgments about the usefulness of AI until you not only have a good deal of hands-on experience, but have seen firsthand how the models and functionality change over time.

The Art of Prompting:
Secrets of the AI Whisperers

This next section serves as a practical guide to help you experiment effectively with prompt writing, a core skill that helps build AIQ. We'll cover the key tactics to quickly get better responses from AI. For more advanced techniques to refine your prompts once you've mastered these basics, you'll find recommendations for additional resources on my website.

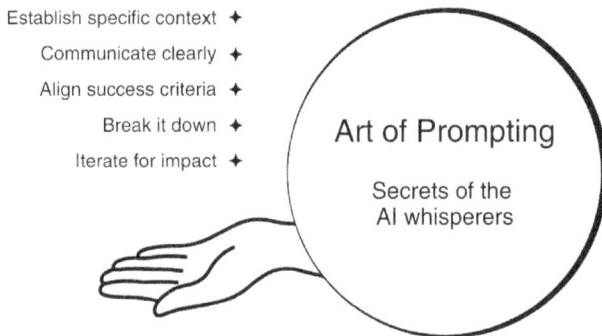

Establish specific context ✦
Communicate clearly ✦
Align success criteria ✦
Break it down ✦
Iterate for impact ✦

Art of Prompting

Secrets of the
AI whisperers

Establish Specific Context

Set the stage with key details that focus AI on what matters most.

As we discussed earlier, AI is a general tool that will provide generic, basic answers until you give it context. The more context you provide, the better responses you will get. For example, include this kind of context in your prompt:

Define your goals: What goals do you have? What are you seeking to achieve?

Assign a role: What should the identity or persona of

the AI be when it does its work? What role do you want the AI to play?

Detail background material: What kind of research do you want your AI to do? Are there documents it should read or charts you are uploading for it to review?

You can give more context and background to "train" the AI in the way you want it to work and the kind of responses you want to see. This could include:

- Uploading style guides, employee manuals, or project briefs.[iv]
- Uploading examples of your writing, or work you are proud of.[v]
- Sharing your relevant expertise. For example, if you're asking AI to help with marketing strategies, you could provide an outline of an effective marketing plan you've used. If you're seeking assistance with personalized meal plans, you might upload a transcript from a podcast interview that discussed nutrition guidance you want to abide by. Or, if you're working on a creative writing project, you could share a framework you've found effective for character development.

Communicate Clearly

Provide direct and unambiguous directions.

Just like humans, today's models can get confused by long, rambling requests. The more directly and clearly you communicate to AI, the better its responses will be.

Be coherent: Organize your thoughts logically and avoid convoluted phrasing.

---

iv   Never upload confidential information into a publicly accessible chatbot.
v   Ditto

Use formatting: Format to help you get your points across. Number steps, clearly label top priorities, put key copy in brackets, and even capitalize words or repeat key points for emphasis.

Express yourself clearly: Use simple words over ambiguous ones. Avoid vague language and be specific about your expectations and desired outcomes.

Be explicit: Steer clear of double negatives—clearly state what to "avoid" instead.

This is a bit of an art, as is all AI prompting. Experiment to find what works best for your specific needs. You'll likely notice that communicating clearly with AI mirrors how we strive to be understood by other people. When refining complex prompts, I often ask myself, "How could I explain this more clearly to a stranger?"

Align Success Criteria

Collaborate with AI to define how it can make its responses more useful to you.

Taking a little time to collaborate with your AI on what defines a "great" response can get you much faster to useful results. Do you want poetry, prose, or code? Do you want quick, billboard-length answers? Do you want answers in a table or an infographic? Is there formatting that will make it easier to make use of its response? Especially when working with AI on recurring tasks, take the time to really align on how its responses are formatted to minimize additional work on your part. Here are approaches to pull from to set clear criteria for success:

Define tone: What kind of response will be most useful to you? Should it be communicated in simple or sophisticated language? Will it be responding with a specific voice or character? Who is your audience? What tone will resonate most with them?

Specify structure: What structure should the AI use to present its responses? Do you have any requirements for how it formats the response?

Outline scope: Are there angles or topics that the AI should cover in its response? Perspectives to consider or avoid?

Set boundaries: What constraints or boundaries do you want the AI to have? Do you want it to be concise? Avoid repetition? Consider the question from only a single perspective? Do you want it to wait for your additional background material before responding?

Consider removing limits: Alternatively, do you want to see how the AI could push your thinking by removing constraints altogether? Asking it to "be creative" and make assumptions along the way can signal AI to take new directions.

Consider putting the AI to work on designing the best way to present its results. Start by providing context and requesting a sample response. Then, engage in a dialogue to refine the format. You might, for example, ask for additional columns in a table or bold text to make it easier for you to see changes. When you've discovered a particularly useful presentation style, ask the AI to describe this approach "in complete detail." Save this description to your library of prompts so you can more easily replicate this approach when you want something similar in the future.

Break It Down

Organize tasks into steps to avoid confusion.
When my kids were little, I learned to break down instructions step by step to increase the chances they wouldn't get confused or miss something. AI also benefits from this approach. Today's AI can stumble when overwhelmed with a barrage of requests, potentially missing details, becoming confused, or executing tasks in the wrong order. When you encounter these issues, try the following:

Break the task into smaller parts: Introduce one step at a time, or explicitly instruct the AI to "think step by step". This technique is particularly useful for tasks that require multi-step reasoning, problem-solving, or complex analysis because it helps the AI find (and combine) the most relevant information more effectively.

Signal with formatting: If a single step has multiple substeps, use formatting to make it more likely the AI will understand you clearly. For example, number steps, or use brackets to identify a specific type of content.

Pace the interaction: Slow it down so it doesn't leap ahead and do something you didn't ask for (and may not even want). This also gives the AI time to "think," which often improves results. Try telling your AI to wait for a response or its next instruction before proceeding.

Verify understanding: Test the AI's understanding of your request by asking it to show its thought process or explain how it will approach a certain task before having it perform the task itself. This can get you to more transparent and verifiable results.

Manage overload: If the AI floods your screen with too much information, constrain the length of its responses by specifying a word limit or a certain number of points it can make.

Ask for help: If you're still not getting the desired results, ask the AI for help improving your prompts. Tell it exactly how its current responses fall short of what you want. Ask it to pose questions that will help it better understand your needs. You can even ask the AI to write a new, improved prompt based on your discussion.

Iterate for Impact

Achieve results through continuous dialogue, refinement, and iteration.

Your initial prompt is just the beginning of a dialogue with AI. Iterate by providing feedback that will guide the AI to return increasingly valuable outputs. This back-and-forth process is critical to unlocking the full potential of AI.

Give clear feedback: Craft your feedback with the same clear communication you used in your initial prompt.

Pinpoint issues: Be thoughtful about what's truly "off" in the AI's response, and share this feedback directly and authentically to help the AI to really understand how to serve you better.

Push it further: Experiment by collaborating through multiple rounds of feedback with your AI. How much better are the responses after half a dozen iterations? A dozen? If you aren't sure what kind of feedback to give, have the AI kickstart your thinking by asking it what would help it deliver a better response, or simply inquire,

"What did we miss?"

Give more context: Consider what additional background material or documents you can provide to help the AI better understand your needs, or give examples of the kind of response that would be more helpful.

Keep tinkering: Remember that small changes in prompting can lead to different results, so continuous experimentation is key. Sometimes techniques such as simply asking AI to "be complete," "say that more clearly," or "make it more readable," or using caps to emphasize a point can improve the conversation.

Reset or reuse: If you're not getting good results, don't hesitate to start fresh with a new chat. On the flip side, when you have a particularly productive conversation, revisit that chat for new tasks that are similar. Alternatively, you can even copy successful chats into a new window and ask the AI to analyze what made the conversation work well; then ask it to write you a starter prompt that captures that effective approach to use in new chats.

With some focused practice, you will soon develop an intuition for prompts that get productive results. Above all, remember: AI already knows how to talk to you, and you already know how to talk to it.

Getting Used to AI Conversations

No matter how digital our lives have become, having an actual conversation with AI can still feel strange at first. Here are examples of prompts my clients have used to start strong and steer conversations toward deeper territory—adapt them to your work, projects, and goals.

Conversation Starters

I am wrestling with a problem and don't know where to begin . . . walk me through an analytical approach to solve it.

I've outlined my initial thoughts, but I know I may have blind spots . . . can you critique my thinking and fill in any gaps?

Our team disagrees on the best path forward. Could you objectively evaluate the pros and cons of each approach?

I need to think through this idea from multiple angles. Can you walk me through it from a financial, operational, and customer point of view?

Conversation "Evolvers"

That response is useful, but seems incomplete. What are we missing?

Take another pass with a different lens (strategic, sustainability, etc.)

Pressure test this against this scenario (economic downturn, evolving customer expectations, etc.)

That suggestion seems too broad. Go deeper with more specific examples and scenarios.

13

Getting to Know AI
Through the Power of Play

Deepen your understanding of
AI's capabilities through purposeful play
with diverse tools and modes of communication.

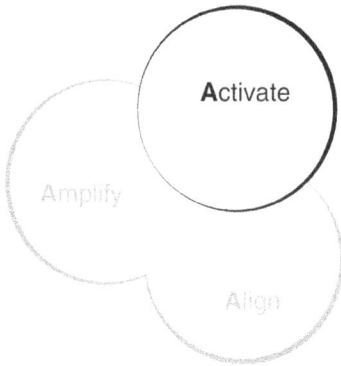

t's time to dive into the first step of the A³ Framework, Activate: a hands-on exploration that will broaden your understanding of the capabilities AI now offers you. And here's the good news: this step is a lot of fun. You'll get to play with a range of tools—and I'll walk you through it all to make it easy.

Adopting this mindset of play triggers creativity and curiosity. It's the perfect way to explore the broad capabilities AI now offers because it encourages experimentation, idea generation, and flexible thinking, all of which are helpful when trying something new.

So why did I open this book with a story about my failed attempt to use play to get my own family to see the value of AI? Because it illustrates that the *way* we play matters.

In my research, I observed a pattern: early adopters frequently explored new tools and releases by toying with them—testing bizarre prompts, crafting unexpected challenges, and inventing quirky tasks just to see what surprises would unfold. And then— critically—it was easy for them to make connections between even strange experiments and "big picture" real-world impact.

However, I found this kind of extrapolation was harder and more time-consuming for newcomers. The key factor for them between fast insight and frustration is **playing with purpose.** I also discovered that just a little guidance makes all the difference. That's why I transformed the unstructured, chaotic play of early adopters into a "PlayLab" that helps people advance productively on their AI journey (while still preserving the creative spirit that makes play so powerful). You'll find all you need on my website in the resources section of ThinkwithAI.org.

Why play?

Research shows that when children play, what appears to be "just having fun" actually functions as a powerful learning mechanism.[57, 58] But kids aren't the only ones who can benefit from this approach—play makes it easier for adults to experiment with ideas, adopt new perspectives, solve problems, and discover new skills. It creates a low-stakes environment in which to take risks and learn from failures, and releases us from holding tightly to preconceptions. And, crucially, by using play to allow the imagination to roam freely we can cultivate the expansive thinking that helps lead to innovation.[59]

This playful approach is especially valuable when you're working to develop AIQ. Because it's so new, understanding how AI can truly benefit you may seem like an abstract, hard-to-grasp concept. There aren't AI user manuals, and even AI experts don't fully grasp all the intricacies of how these systems actually work. By diving into hands-on play, you can bypass the need for concrete how-tos and drop directly into the experimentation that will cultivate an *instinct* for working with AI.

This is a whole lot different from how we typically engage with software, right?

Let's Play: Introducing the PlayLab

Ready to get your hands on AI? Head over to my website, at thinkwithai.org/playlab, where you'll find the PlayLab, a collection of playful challenges that give structure and purpose to your exploration. Choose your adventure based on how much time you have (I have challenge sets starting at just twenty minutes) and whether you're sticking with free tools or ready to invest a little. (The price of a pizza or two unlocks a world of AI possibilities, and believe me, it's worth the investment.)

Pick a challenge set that intrigues you and jump in. I assure you that you will walk away with a broader appreciation of what today's AI can do.

Fair warning: you might encounter some bizarre or laughable AI responses (what I affectionately call "AI bloopers"). But don't be discouraged! These quirky moments are part of the journey. And who knows? After a few duds, you might stumble upon a surprisingly brilliant answer. When that happens, pause and reflect—it could be the gateway to your own "aha" moment about

AI's potential.

As you explore the new capabilities AI offers, you will also get to experience:

How to "talk" with AI in multiple ways: AI is multimodal. It can see you, hear you, and even talk to you. We can interface with AI by typing, through a chat box in a browser, or through an app in which it's been integrated. We can have a conversation with it using our voice, and choose a voice for it to use to talk back at us. We can give it a photo of our dog or drawings on a whiteboard and ask it to do a task based on these images. AI can even draw for you, turning sentences into a chart or a picture. You'll get a chance to experiment with all these modes in the PlayLab.

How AI handles open-ended goals and creative tasks: AI can perform a broad range of work and create or even "imagine" in a way no software you've ever met before can. It can tackle open-ended challenges and get creative in ways that might surprise you. Dream up any category of work, and chances are AI can contribute meaningfully. In the PlayLab, I show you some fun examples to spark your imagination, but remember, we're just scratching the surface.

The PlayLab offers a fast, focused flyover of various AI tools, capabilities, and ways to communicate. However, it won't demonstrate how these initial experiments translate into lasting value in your life. That's why the next chapter shifts our focus from play to practical application, helping you pinpoint how AI capabilities can connect to your unique demands and workflows. This next step of the journey, "Align," is designed to guide you to an "aha" moment that reveals how this technology can genuinely serve *your* needs.

14

Step Two: Align

Get AI to Truly Serve You

Identify where and how AI can be meaningful to you, and run focused experiments to discover a specific breakthrough "aha" use case.

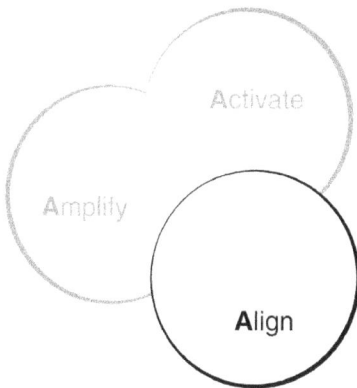

Now that you've got your hands in AI and have a feel for the kinds of things it does, it's time to discover how AI can truly serve *you*. The goal is to get to an "aha" moment—a real, bona fide use case of how AI can genuinely enhance your life in a meaningful way. This work consists of two stages. The first is to

Focus your attention on a specific challenge; the second is to *Experiment* with solutions to this challenge:

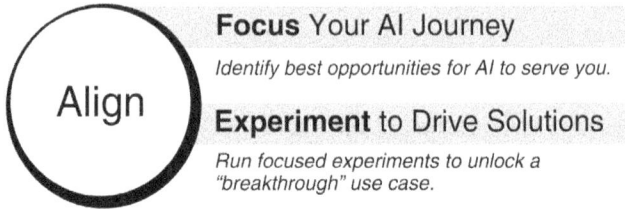

Focus Your AI Journey

Identify best opportunities for AI to serve you.

Align

Experiment to Drive Solutions

Run focused experiments to unlock a "breakthrough" use case.

This step transforms the essential early adopter practice of rapid experimentation into a structured process that points you directly to valuable use cases. To align your needs to AI's capabilities, you'll embark on a thoughtful *discovery* process that reveals opportunities unique to you. It demands some deep thinking about what would help you to take the things you do (whether personal or professional) to the next level. I know this emphasis on self-reflection can feel strange—we are talking about working with software, after all. Even the language I use here may feel surprising at first. But stick with me here: this focused approach drives you much faster to value.

From Fixed Functions to Fluid Intelligence

Traditional software helps us with well-defined tasks that need consistent, predictable answers. We compile sales figures using reporting tools, track project deadlines using project management tools, and analyze website traffic patterns using analytics tools.

In contrast, because AI is a broad intelligence trained on a vast corpus of human knowledge, it can help us with open-ended goals and even loose objectives in virtually any area. It can interface with us at different levels of sophistication—from simple kindergarten-level language to something that looks as if it could come from someone with a PhD.

This open-ended opportunity is exciting, but it's also why many find it hard to use AI well. We're not used to deeply contemplating where and how to apply our software, which is a big reason some people get frustrated and quit before they see how useful AI can really be. But when you start by *really* considering your true needs, you're setting a foundation to discover how AI can bring you real and important benefits in any role you play, from career to family life, regardless of your profession or personal goals.

A Personalized Journey to Success

We're all arriving at this moment with our own unique blend of strengths and weaknesses. Each of our lives is filled with its own kaleidoscope of demands, goals, and responsibilities. And AI is always evolving, so it's hard to predict where it works well, and where it doesn't. These are all reasons why no one can hand you a template or point you to a user manual that will tell you how to best use AI.

While there is plenty of inspiration to be found in others' successes, achieving results is a unique journey for each person or team—and practice and experimentation are the most effective ways to drive a win.

The process I walk you through in this chapter guides you in finding a match between your needs and AI's capabilities—and develops your skill in tapping into those capabilities.

This process is designed to deliver tangible results quickly. That said, you may find you don't need to go through all these steps and after a quick scan feel ready to dive into the "Experiment" stage. If that's you, please do—the sooner you start experimenting with AI to address real needs, the faster you'll see results. However, for those less comfortable with AI tools, taking time to fully go through each part of the process can be beneficial.

Focus Your AI Journey

The first stage is to focus your efforts on addressing real needs—areas where AI assistance can significantly impact your life or help you overcome barriers to higher performance. This section takes you through a three-part process to sharpen your focus. The more introspective and thorough you are in this approach, the more targeted and productive your AI experiments will be.

Focus Your AI Journey

Align

A. Identify Your Needs

B. Choose Your First AI Challenge

C. Imagine Your Ideal Cognitive Partner

Starting with real needs serves an additional purpose. It raises the bar and sets higher expectations for your interactions with AI. This mindset shift can help overcome preconceptions that might limit your exploration. By setting ambitious goals for what AI can deliver, you'll be motivated to push harder, increasing your chances of achieving a meaningful breakthrough. Remember, the more you demand from the technology, the more likely you are to uncover its true potential and value for your unique situation.

I've found that those who are most successful with this discovery process carve out time in an environment where they can think expansively. They're capturing their thoughts without self-censorship or judgment. Some even find it helpful to journal to uncover deeper insights.

A. Identify Your Needs

First, allocate focused time for brainstorming and begin compiling a list of specific needs, challenges, or aspirations that could potentially benefit from AI assistance. Look beyond current pain points to include future goals and aspirations. Consider any areas where you could benefit from enhanced capabilities, efficiency gains, or a cognitive boost.

To guide your exploration, ask yourself these questions:

- What are your unique strengths, and what would help you amplify or leverage these strengths further?
- Which tasks or responsibilities consistently drain your energy or frustrate you?
- With which personal or professional goals have you struggled to make progress?
- When and where do you feel blocked?
- Where do you experience bottlenecks or feel limited by your current resources, capabilities, or knowledge?
- What aspirations or dreams have you put on hold due to constraints or barriers?

B. Choose Your First AI Challenge

Now that you've compiled a list of potential areas from which to explore AI's value, it's time to prioritize and select a single challenge to focus on. While it may be tempting to tackle multiple needs, concentrating on one specific area allows for a deeper discovery process and increases your chances of achieving a real breakthrough. After succeeding with one use case, you can return to reuse the process and tackle more needs.

Choose your mission carefully: you'll be investing considerable time and effort into cracking whichever challenge you select. The point is not only to get something truly useful from this initial deep dive, but also to gain experience and learn a process that you can keep repeating to unlock more value over time.

I encourage you to choose a challenge slightly beyond what you expect AI to handle. While basic administrative help is useful, try to go for something that will require finding a way to have AI really "think" *with* you ... something that will demonstrate its "creative" and "strategic" side.

If a specific need or aspiration from your list strongly resonates with you, trust your instincts and proceed to the next part. However, if you're having trouble deciding, evaluate your options across these dimensions—the one that scores most highly will be the one most likely to yield valuable results.

Evaluating Your AI Challenge

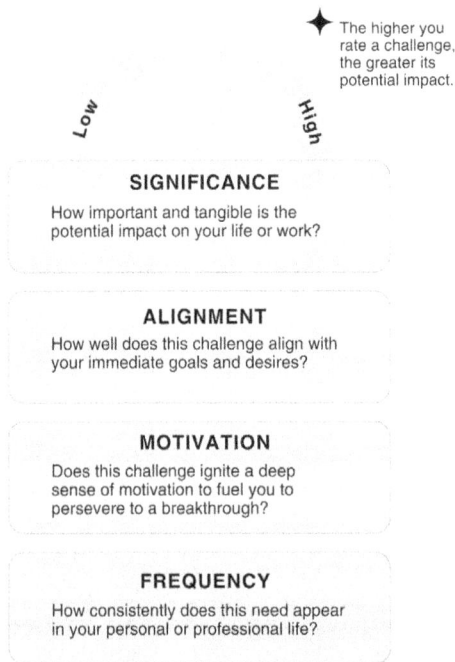

The higher you rate a challenge, the greater its potential impact.

Low   High

SIGNIFICANCE

How important and tangible is the potential impact on your life or work?

ALIGNMENT

How well does this challenge align with your immediate goals and desires?

MOTIVATION

Does this challenge ignite a deep sense of motivation to fuel you to persevere to a breakthrough?

FREQUENCY

How consistently does this need appear in your personal or professional life?

Once you've chosen your challenge, focus your work by framing it as an AI Challenge using this template:

How can AI help me . . . _____ ?

[specific goal/need]

for example:

. . .   develop a comprehensive business
         plan for my startup idea?

. . .   overcome writer's block to get to a
         compelling narrative for my novel?"

. . .   put lifestyle changes in place
         that are actually sustainable?

. . .   uncover insights for my doctoral thesis
         on sustainable urban development?"

. . .   understand advanced biology
         concepts more effectively?

C. Imagine Your Ideal Cognitive Partner

Before diving into the experimentation stage, I've found it's useful for people to take a moment to first deliberately stretch their imagination about the role AI can play. I encourage you to temporarily set aside the fact that this is a book about AI, and visualize a hypothetical *human* collaborator that has the ideal blend of skills, expertise, and traits to address your specific need. I know this exercise can feel unconventional or even silly, but deliberately starting your thinking from "human help" can push your expectations of what software can do for you.

Imagine a versatile, highly capable partner who can think creatively, strategize, and provide support that's uniquely tailored to your needs. Consider the following questions:

- What specific roles or responsibilities would they fill to help you overcome your challenge?
- What skills would they need—or what knowledge domains must they possess—to be truly valuable to you?
- How would they amplify your strengths or compensate for your weaknesses or knowledge gaps?
- What unique perspectives or approaches would be helpful for them to bring to the table?

Use what you learned to add more texture to your AI Challenge by following this template:

How can AI serve as a _____
                          [role/capacity]

to help me _____ by _____ ?
           [specific goal/need]   [specific task]

Consider getting even more targeted by breaking your challenge into sub-challenges, each representing a single component of the work. Remember, the more deeply you understand your AI Challenge, the more likely you will be to achieve a successful breakthrough.

For example, you could break down the AI Challenge "How could AI help me to finally launch a successful vlog?" into more specific AI Challenges such as:

"... define my niche and audience,"

"... create a name and brand,"

"...develop compelling content and a publishing schedule," or "...promote my vlog."

Then, imagine the ideal cognitive partner for each component. To illustrate, here are a few more specific AI Challenges addressing different content needs for our vlog example:

How can AI serve as a . . .	to help me . . .	by . . .
content strategist	generate engaging topics for my vlog on sleep	analyzing trends, summarizing studies, and giving me insights into audience interests?
content developer	develop compelling scripts for my vlog episodes	honing my tone and voice, and making sure I'm clear?
content optimization specialist	increase the visibility of my vlog	suggesting SEO keywords, optimizing metadata, and enhancing visuals?

Pro tip: get help from AI through this entire process. Simply select a tool, describe your challenge, and start a dialogue on what's hard about it for you. Ask AI to help you break your challenge down into smaller parts. Have it brainstorm what an ideal cognitive partner would be like. You can even throw it the list of questions I asked you at the beginning of Part C. The more specific you are, the more targeted and useful the AI's responses will be.

> I am launching a vlog on the importance of sleep. My goal is to achieve a high number of followers. I am:
> 1) struggling to get content development started and
> 2) I don't know how to set it up for success.
> I want to bring on a cognitive partner to help me, and need your help in helping me to visualize the roles and responsibilities that an ideal partner would play. Start by:
> A) describing "a perfect partner" and
> B) brainstorm what specific things they could do to help me.
> Be comprehensive and specific.

You could even experiment with asking open-ended questions to expand your thinking. For example, copy any notes or drafts you have into an AI chatbot and ask, "What other questions should I be asking myself to help me visualize this person?"

Visualizing the perfect cognitive companion is a powerful exercise to bring the support you need to life. You now have a focal point to guide your experimentation to get AI to truly serve you.

Experiment to Drive Solutions

In this next stage, Experiment, you will actively engage with AI to find solutions to your AI Challenge. Your goal is to identify your first breakthrough use case—an "aha" discovery of how AI can become an indispensable tool in your life.

People often ask me for "the" prompt to get to a breakthrough. While there are some tried-and-true places to start for many common challenges, there is not a one-size-fits-all roadmap to breakthroughs. This is because everyone's needs are unique. That's why this stage is about *you dedicating time free from distractions for freeform exploration.*

So set aside some time and take the plunge to go explore! Your AI Challenge will be your guiding star as you experiment. Put this on a sticky note by your monitor, write it on your whiteboard, print it out and put it on your fridge, or whatever else you typically do to keep a thought front and center.

Work through a creative and iterative collaboration with AI to uncover results, a journey that demands patience, persistence, and an open mind. While it's personal and often unpredictable, there are some basic principles that can guide you to success:

Experiment with a range of tools. Try meeting your AI Challenge by returning to the AI tools you played with in Step One: Activate, and experiment with a few different tools. Also, research whether there are AI tools and solutions that were designed specifically to help with your particular need. If these domain-specific tools are available, give them a try, but also run your challenges through at least two frontier model AI chatbots.

Use good principles of AI-human communication. Interact with the tools following the principles of AI-human communications outlined in Chapter Twelve.

Don't reinvent the wheel. Research how others have tackled AI challenges similar to yours, especially when you encounter obstacles. There are people all over the world experimenting with using AI for virtually every conceivable application, and it's likely someone has already discovered, tested, and documented their use of AI for a challenge similar to yours. Search the internet to find relevant videos, tutorials, and guides.

Lean heavily on AI at every step along the way. Leverage AI as a collaborator throughout your process, from initial brainstorming to prompt crafting and analyzing results. Experiment with AI assistance when you're unsure of next steps or want to test or evolve your thinking; you can even ask AI to do the work of a particular step for you. For example, start a chatbot conversation by throwing it your AI Challenge question verbatim.

> You are an expert in crafting effective AI prompts. What key principles should I follow to write prompts that help AI generate a successful social media strategy for promoting my vlog across different platforms?

If you like the responses, ask the AI to write a prompt for you.

> Write a highly effective prompt based on these principles.

Turn to AI to continue to brainstorm ideas, suggest your next steps, conduct research, or analyze results. Simply go to a chatbot and "think out loud" by telling the AI why you are stuck and asking for suggestions to help you break through.

Set aside a small budget. The free versions of publicly available chatbots will absolutely get you started. But I highly encourage you to set aside at least a modest budget to try out the more advanced models and robust functionality or give you access to domain-specific tools. Without a little bit of spend, you are not going to see how AI is truly capable of helping you.

Persist. If you aren't getting good results from the tools one day, try a different approach, a different tool, or even revisit the work on another day. Achieving tangible results requires focused and deliberate practice. You might find your first breakthrough in ten minutes, or it might take you a few hours. I think of this investment as similar to when a new team member joins your organization. There is a period in which you are gauging their capabilities, and you work together to discover how they can best support you to achieve your goals. But it typically takes some initial time and patience to establish a productive working relationship. The same principle applies here.

While humans are still better at most tasks, AI is rapidly advancing and now matches or surpasses the average person's performance in a growing number of areas. As you experiment with solving your AI Challenge, it's critical to understand that AI has a "jagged frontier." This means that there are some things it's very good at and some things that it's just terrible at—and it's very difficult to accurately predict on which side a particular task will fall.

That's why it will be extremely important to explore various facets of your challenge to uncover where AI truly shines. In some cases, AI might help you attack your entire challenge, and in others, it may only be able to support you by helping with a small task. Only through trial and error will you discover which aspects of your AI Challenge can be effectively addressed by AI—the technology might excel in one area while falling short in another. Note that there are some AI Challenges that may just not be a fit with the tools right now—they simply aren't good enough yet in some areas. In this case, go back and grab another challenge from your list and try again.

However, by the time you've got through a couple of experiments and have had some substantive dialogues with AI, you'll likely have found at least a spark or a glimmer of what today's tools can do for you. Keep pushing, exploring, and experimenting—always keeping your focused AI Challenge at the forefront of your mind—until you achieve a breakthrough.

You'll know you've hit on something truly valuable if, once you've had a breakthrough, you don't want to give it up—if your AI were shut off, you'd feel like you would *need* it back.

You Got to a Breakthrough! Now What?

First of all, congratulations—this is a key milestone in developing AIQ! Now, take a moment to reflect on your experience. Consider both what led to the inspiration for this breakthrough and any concerns that arose. Think about unexpected insights or novel approaches that emerged and how this breakthrough might apply to other areas of your life or work. Reflect on moments where the AI sparked new ideas, and imagine future possibilities this could lead to. What lessons can you apply to future AI Challenges to achieve quicker breakthroughs next time? Are there ethical considerations you need to explore before applying or leveraging the results? Could your breakthrough benefit others, and if so, how might you share it effectively?

Next, document your success. Start a file with a copy of the effective prompts, and if possible, preserve the conversation in your AI tool. Label it clearly so it's easy to continue your work in the future.

In our next chapter, we'll go deeper into how to leverage this breakthrough to get even further in the quest to make AI serve *you*.

15

Step Three: Amplify

Beyond the Breakthrough

**Build on your initial breakthrough to create a
cycle of continuous improvement and discovery,
transforming AI from a single-use tool into
an integral part of your daily life and work.**

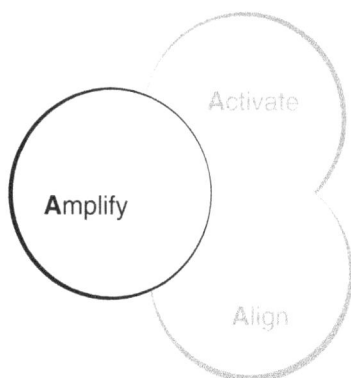

Your first AI breakthrough isn't an endpoint—it's a launch pad for even greater possibilities. Here, we explore how to amplify its impact. I've identified two powerful approaches that can help you transform a single breakthrough into sustained performance gains: *Expand* and hone your methods, and *Integrate* AI seamlessly into your daily life and work.

Amplify

Expand Success
Evolve methods to continuously expand value.

Integrate AI into Your Life
Weave AI into your daily routine.

These two stages work best when pursued in parallel. They have the most impact when viewed as ongoing initiatives, which helps you incorporate AI into your life while keeping you connected to the evolving capabilities of AI models.

Expand Success

There are three specific steps that will help you to *Expand* the value you derive from your breakthrough.

Expand Success

Amplify

A. Refine Your AI Collaboration

B. Streamline for Repeat Success

C. Repeat for New Discoveries

A. Refine Your AI Collaboration

Get better at extracting value from AI in service of your AI Challenge with these three refinements.

Hone your prompt: Whatever prompt or series of prompts got you to your breakthrough, don't consider it fully baked—ever. Just like a chef continually tinkers with a recipe, continue to experiment and refine your prompts to see what gets you better results and how different prompts work in different tools. Also, because the tools are constantly changing, you may find a prompt that was working well suddenly returns low-quality results. Don't assume this is definitive—it may change again. Revisit tools— don't write them off after one bad experience.

One of my favorite tactics is to get AI involved in honing my prompts. Simply tell it your challenge, your current prompt, and some thoughts on what kind of results would be more valuable to you—and ask it to give you suggestions to get you better results. Some software providers are even baking this kind of prompt help right into the tools.

But it's also helpful to revisit your research on a regular basis to see whether others have made and shared new approaches to solve AI Challenges that are similar to yours. As you know, this is a learning moment for everyone, which also means that there are a lot of people continuing to find and share their successes along the way.

Develop your data: Data is a basic fuel of AI. By being thoughtful about additional data that could help AI help you, you could get better results. The models are already operating with the benefits of the vast data they hoovered up in their training. Ask yourself: what data can you access that could help AI deliver better results for you? For example:

Task	Data That Could Help AI Help You
Draft social media posts in my unique tone	Examples of your own writing, social media engagement metrics, successful past posts, brand guidelines, target audience profiles, and competitor analysis
Develop easy start guides and customer service scripts for my product launch	Product manuals, descriptions of customer personas, meeting notes, beta testing feedback, FAQs, competitor product guides, and user journey maps
Identify product improvements and prioritize feature development	User feedback data, usage analytics, customer support tickets, market research reports, competitor feature lists, and industry trend analyses
Create a personalized nutrition plan	Food diary, health metrics (weight, blood pressure, etc.), fitness goals, allergies/intolerances, food preferences, and favorite recipes
Design a custom garden plan	Soil test results, sunlight exposure data, local climate information, plant preferences, yard dimensions, water availability, and photos of the space

AI chatbots can ingest quite a bit of background explanation, files, templates, and other forms of data to guide them in developing responses. This data and guidance that you input into AI is your "context window," but you can think of this as the immediate memory of the system. Context windows have been rapidly expanding so that you can input the equivalent of up to hundreds of pages of text (and even more is available to corporate customers), which makes it possible for AI to review entire technical documents, long literary works, and lengthy financial reports.

For many tasks, it's certainly not necessary to use that much data, and as of this writing, the models can get "confused" at times from large amounts of content. Breaking the work into smaller tasks and collaborating with the AI step by step goes a long way to keeping the quality of responses high as you work with more data.

Sometimes it's valuable to take an extra step and have the AI synthesize data into a guide or template that can be quickly uploaded in future AI chats. These "Reusable AI Briefs" give you a moment to review and check AI's "thinking," collaborate on refinements, and provide an efficient method for guiding your AI in future interactions.

Examples of Reusable AI Briefs

Create a personalized style guide by providing writing samples, so AI can consistently emulate your voice in future writing tasks.

Develop a meeting best practices guide by analyzing meeting transcripts and notes, which you can then use to improve agenda creation and meeting materials.

Generate a custom productivity blueprint by reviewing articles on your favorite time management techniques for analysis of your calendar and project plans.

Retest: Regularly test your AI Challenge across different tools and with new releases. AI models evolve rapidly, and your preferred tool today may be outperformed next quarter (or next week!). Stay proactive by experimenting with each new release to see whether it yields better results for your specific needs. This ongoing evaluation ensures you're always leveraging the most effective AI solution for your challenge.

Don't Feed the AI: Keep Your Private Data Out of LLMs

It's very important that you are not giving the Large Language Models (LLMs) any data that is private or proprietary. Some LLM providers explicitly state that they may use your inputs to train and improve their models, and anything you paste into a chatbot—whether financial data or customer communications—is essentially outside of your or your company's control. As a general rule: never input non-public corporate information into public LLMs, consult your organization's policies before using AI for work-related tasks, and collaborate with your IT department to ensure you're operating in a safe environment.

B. Streamline for Repeat Success

Here are my favorite tactics to streamline a breakthrough so it's easier and faster to execute it again.

Capture your findings: Save your effective prompts and instructions in a file for easy access and sharing with others. Believe me, when you're moving fast, they quickly get scattered or lost, and something as simple as having quick access to your prompts makes it more likely you will continue to use them. This can be as simple as a Microsoft Word document, a Google Document, or an Airtable for easy sharing, or it can be a more involved database. I often use Notion to help me organize my work, with different sections for AI Challenges in different stages of development and links to my Reusable AI Briefings.

This step is especially important for teams, where you may have multiple people working simultaneously to identify and find solutions to AI Challenges. To address this need, a new category of tools is emerging: prompt management platforms. These platforms help teams centralize and organize prompts, track iterations, share successful strategies, ensure consistency, and maintain version control.

You can also use the memory function some chatbots have to help you. Depending on the model and functionality, there are chatbots that can "remember" what was said previously in a chat and, in some cases, retain information on preferences across different conversations. This means you can build on a conversation, adding to it or asking it to go back and change something without having to start over or reintroduce a topic. Or if goes off track, tell it to go back to the point where it was working better.

If you get AI going in a way that is working for you, return to that same chat another day and you can start from that point in the conversation. Or copy and paste a previous chat into a new one and ask AI to analyze the past conversation, pulling out the decisions made, the tone that was used, and how responses were presented—and to create a prompt that can be used to leverage these elements for new work. I keep a set of these analyses handy in my library so that I can choose from different tones and kinds of interactions as a starting point for future tasks.

Automate repetitive tasks: Many providers offer functionality to save and reuse instructions, even allowing you to share these customized setups with others. Different tools have different names, but the core concept is consistent: making it easy to reuse personalized AI instructions to streamline your work.

ChatGPT, for instance, enables you to create a dedicated AI "bot" trained on your specific data and instructions. (Called

"GPTs," these are basically custom versions of ChatGPT.) This process is surprisingly straightforward, as you can "code" instructions using natural language without any programming knowledge. Creating a useful GPT can take just minutes. Simply provide clear instructions on desired tasks, necessary context or data, and any specific requirements. (I suggest documenting all instructions and notes in a separate file outside of ChatGPT as a backup.)

Anthropic's Claude has functionality called "Projects" that allows you to set up a project space that can be used to customize the chatbot experience and bring more efficiency to repeated tasks or work that requires the same background data and instructions.

A "lighter" tactic I often employ within a chat is creating code words that trigger specific response styles. When the AI nails exactly what I need—be it tone, detail level, or format—I ask it to describe its approach and assign a short acronym or word as shorthand to "do the same thing" in the future. Over time, I might accumulate a half dozen codes in a single chat, each activating a distinct type of response with just a few keystrokes. If I want to take it into a new chat, I simply paste the write-up that the AI created for me into that new chat. It can feel like we've developed our own secret language to help us work together as a faster and more effective team.

Connect to other apps: You can get more value from AI for some tasks if you connect the tools to your data inputs, or connect the AI's outputs to other software which then can take action on the results. For example, output from an AI analysis could automatically create tasks in your project management software or send notifications to your team-messaging app. Various platforms offer different levels of AI integration and automation capabilities. For instance, some make it easy to trigger actions in one app based on events in another, or automatically feed data from multiple sources into your AI system. While some of this work currently requires technical expertise, vendors are actively working to simplify these processes. We can expect significant advancements in this area, with both entrepreneurs and established platforms striving to make AI integration increasingly accessible and user-friendly.

C. Repeat for New Discoveries

I encourage you to leverage your learning investment as a springboard for further discoveries. Simply repeat Step Two, "Align," to continuously identify new needs and new AI Challenges. Over time, you can build a library of breakthroughs that support many areas of your personal and professional work.

You can also find inspiration by following people in the AI community who are sharing their own discoveries. (I've included some recommendations on my website.) Make sure to check out OpenAI's GPT "store" in which people from all over the world make their GPTs available to the public. It's a treasure trove of ideas, and I consider it a global laboratory of innovation, where we all can peek into how other people are currently thinking about using the technology.

The journey of learning how to leverage AI is not a one-time event. It's a continuous process of exploration and innovation. I encourage you to keep identifying new needs and developing new AI Challenges. Over time, you'll build a comprehensive library of breakthroughs that support various aspects of your personal and professional life. Each discovery becomes a springboard for further insights, creating a compounding effect on your AI proficiency and its impact on you, your creativity, and ultimately, your performance.

Integrate AI into Your Life

Once you see how AI tools can help you, the next step is to *Integrate* them as a regular part of your life so you are continually getting valuable support from AI. Here are three next steps to continue to get returns from the learning investment you have made so far:

Integrate AI into Your Life

Amplify

A. Develop an AI Habit

B. Bring AI into Your Routine

C. Make it a Practice to Tap in

A. Develop an AI Habit

You've gone your whole life without this kind of AI. You've developed all kinds of engrained habits and approaches to your work that don't include using AI. And so now, don't be surprised if one of the biggest challenges you have getting going in AI is even remembering that AI is a tool you can turn to. However, the same approach that can work for incorporating exercise and healthy eating into our daily lives is also effective for amplifying AI's value to you.

The author James Clear, in his book *Atomic Habits*, speaks to a strategy of building habits through small, consistent actions. He applies a "two-minute rule." To develop a new habit, find a way to perform that new habit in just two minutes. "Take whatever habit you're trying to create and scale it down to something that takes two minutes or less to do," Clear explained. "So 'read thirty books a year' becomes 'read one page' or 'do yoga four days a week' means 'take out my yoga mat.'" The secret is to always stay below the point where it feels like work. Clear calls these "gateway habits" which can make a targeted new habit feel less daunting and more achievable, emphasizing that with new habits, the goal is consistency, not mastery.[60]

Even small changes such as taking the stairs instead of the elevator or drinking more water require this kind of attention to turn into automatic behaviors. Turning to AI requires the same kind of intentional effort and diligence. Many people have shared with me that they were putting off using the tools until they "had time to really dig in." Dedicated time is hard to come by in our busy, modern lives. Instead of pushing off exploration until the mythical "later" that may never come, I tell people to apply the two-minute rule instead and keep it up for a month. They invariably come out of that month with a lot of energy and excitement about what they've found, and often have learned to consult these new thought machines as naturally as they check their smartphone.

The great news is that two minutes can get you pretty far in AI. It's enough time to think of one thing you want help with, develop a quick prompt, and then run that prompt through a tool or two and give the results a quick look to see if they are helpful. Do a different kind of prompt experiment every day for a month, and you'll have a solid start to your discovery process. Over time, turning to AI for help will become as natural a habit as turning to a search engine.

B. Bring AI into Your Routine

Take your use of AI further by practicing ways to bring it more actively into your day. Here are three easy-to-implement recommendations to deepen your practice of turning to AI.

The Daily Question: At the start of every day, ask yourself "How can AI help me today?" Be curious about whether and how it can help you with something, even if you aren't sure it's capable yet. I keep a list in my notes app on my phone so I always have a starting point when I want to run a new experiment, even if I don't have time to work on it that day. I've found this process helps people expand their thinking about where and how they can use AI to get a boost.

The AI Bake-Off: Let me share one of my most effective strategies for staying at the front edge of AI while building your daily practice: what I call the AI "bake-off"—a simple but powerful approach to testing multiple AI tools against each other. Pitting the tools against each other continually gives you a front-row seat to see their evolving strengths and weaknesses up close, and helps you quickly identify which tool is best for what kind of tasks.

The best part is that it adds mere seconds to incorporate an AI bake-off into your routine. Pick your favorite two to three tools and keep a browser tab open for each. When you hit a struggle in your work—perhaps you are grappling with writing a sentence or encounter a concept you don't understand—throw it to the tools. Put in a prompt to one, copy it, and while it's running, hit one or two more with the same or similar prompt. It's typically easy to see which is returning the best results, and that's the one I focus on for each particular task, ignoring the rest until the next bake-off.

By running the same prompts through multiple models simultaneously, you are continually assessing which tool performs best for your particular needs at any given time. The models are advancing rapidly, so their relative capabilities can shift from one day to the next.

Companies are doing bake-offs too. While initially,

corporations may have been experimenting with a single model or two, now many are testing multiple models so that they can be thoughtful about which use cases they channel to which model (for example, weighing cost versus performance for a specific use case) and to tap into changes as the whole space continues to move rapidly.

The Open Invite: Invite AI to the table for every brainstorm. This could be personal—like figuring out where to take your next vacation—or professional, like working on the next quarter's marketing campaigns. It's a great way to discover more value from AI as a thought partner. During brainstorms, I've found it valuable to have AI running on a side screen, continuously processing our discussion according to my pre-set instructions. It's like having an ever-present, low-key participant who's always ready with ideas. While we might only use a small fraction of what it generates, this constant stream of AI-produced thoughts can significantly enrich our ideation. It's there when we need a quick glance for inspiration or a fresh perspective, offering possibilities without interrupting the flow of human conversation.[i]

---

i   You'll find more on how to leverage AI as a team in Chapter Sixteen.

C. Make It a Practice to Tap into the AI Pulse

The AI space is vibrant with innovation and critical discussions on responsible AI, but it's hard to know where to start because there are so many people talking about the space and so much content to wade through. To help you tap into AI and engage with the latest innovations, insights, and discussions, I've provided a curated list of resources that I've found to be particularly effective. I make it a practice to "time box" a segment of each day for catching up on new developments or experimenting with new tools and approaches. However, for most people, a weekly update is probably enough. As you develop familiarity with these learning communities, you'll find that you can get a lot of content even if you can devote only a few minutes to this kind of regular practice.

The A³ Framework:
Your Foundation for Continuous Learning

AI is always evolving. Your needs are ever-changing. This is why the journey to getting AI to better serve you never really ends.

It's not about reaching a destination—it's about keeping pace with possibility.

Now that you are familiar with all three steps of the A³ Framework, you are better equipped to keep pace with possibility. Through purposeful Activation, you've expanded your understanding of AI's potential. Through strategic Alignment, you've matched AI capabilities to your specific needs. And through consistent Amplification—including the regular learning habits we just discussed—you've learned how to integrate AI into your life. This progression has moved you closer toward the AI Boosted zone, the highest level of AIQ, where you intuitively engage with AI as a partner to enhance your creativity and performance.

Levels of AIQ Progress

AI Curious	Starting the journey

Beginning to explore AI, experimenting with its features but not yet using regularly or for specific outcomes.

AI Collaborator	Building momentum

Identified valuable, practical uses for AI in specific tasks, and starting to see AI as a tool to extend creativity + productivity.

AI Boosted	Achieving leverage

Integrates AI naturally as a thought partner supporting even strategic tasks to boost performance and creativity.

The A^3 Framework is an enduring foundation to help you systematically explore and integrate emerging AI developments as they appear. As you continue to develop your personal AIQ, you're also uniquely positioned to guide others on their journey. This brings us to our next critical challenge: how to elevate the collective AIQ of a team.

16

Enhancing Team
Performance by Thinking with AI

Companies are exploring diverse strategies to harness AI's potential. Some are racing to understand how adding AI to products or services could make them more useful to customers. Others are redesigning business processes to delegate entire tasks or even workflows to AI. But as we've seen throughout this book, people across every part of the organization could also pull AI in dynamically throughout their day to boost their performance and creativity across a wide range of work.

Using AI responsibly at work involves complex considerations around governance, privacy, security, and potential bias. Deep exploration of these topics is beyond this book's scope, but this chapter provides a brief introduction to how to help teams advance on their learning journey to use AI as a collaborator and thought partner.

When teams learn to Think with AI together, they don't just understand the technology better—they develop collective intelligence that helps them spot opportunities, avoid pitfalls, and drive innovation more effectively. This collaborative approach not only boosts performance but helps to build the organization's AI capabilities—all while maintaining heavy human oversight.

Even with this oversight, however, it is critical to coordinate with teams responsible for AI governance and compliance from the beginning.

To help get you started, I'll walk you through some of the ways I've seen teams effectively use AI in their work. Then, I'll share some approaches that I've found particularly effective for groups learning to collaborate with AI.

How Can AI Serve Your Team?

Here are some of the ways AI can support team performance:

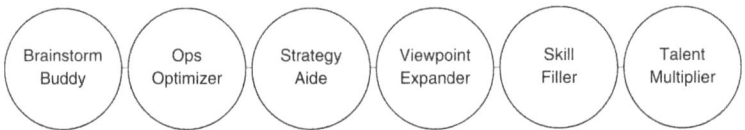

Brainstorm Buddy	Ops Optimizer	Strategy Aide	Viewpoint Expander	Skill Filler	Talent Multiplier

Brainstorm Buddy: Inviting AI to brainstorms can give your team's collective imagination a turbo boost. Before a brainstorm, AI can analyze vast amounts of data and generate key questions to kickstart the conversation. During the session, set up a side screen with AI ready to chime in with ideas or answer questions as they arise. It can help evaluate concepts from multiple perspectives, provoke new lines of thinking with unexpected questions, or offer those wild ideas that humans might hesitate to voice. AI can also provide structure, ensuring all aspects of a problem are considered, while offering rapid feedback to refine ideas. It can synthesize a discussion or transform whiteboard notes into visual infographics.

Ops Optimizer: Identify areas where AI can serve as operational support for your team's routine tasks. By codifying your team's standards and best practices into AI tools,[i] you create always-ready assistants to develop materials or analyze and synthesize content, helping your team move work forward efficiently. Teams have used this approach to run marketing materials through compliance checks, draft project briefs, identify key takeaways and next steps from meetings, create optimized meeting agendas, and prepare quarterly update slides.[ii] This AI-powered operational support frees up human team members to focus on higher-level tasks while maintaining consistency and efficiency in routine work.

Strategy Aide: Your team can leverage AI to support strategic thinking and planning processes. By integrating AI into high-level tasks, you can accelerate decision-making, enhance analysis, and uncover angles your team might overlook. AI can assist in conducting SWOT analyses, developing risk assessments and running simulations, drafting crisis management plans, evaluating brand positioning strategies, generating novel product concepts, and performing in-depth market analyses—and just about anything else.[iii] When used responsibly, AI allows your team to approach strategic work with a broader perspective and more robust data-driven insights to support informed and innovative strategies.

Viewpoint Expander: Teams can leverage AI to expand

---

i   Many providers have features that allow you to save and reuse customized instructions across a team. For example, ChatGPT allows you to create custom "GPTs" that follow your specific guidelines and data. Claude has self-contained workspaces that include their own chat history and knowledge bases called "Projects." Approaches such as these require no programming knowledge and can be set up in minutes, helping teams to develop repeatable workflows.

ii   You can enhance the value of AI by connecting it to other software, allowing for automated actions such as creating tasks or sending notifications. While some integrations still require technical expertise, many platforms are making it easier to link AI with various tools, and significant advancements in user-friendly AI automation are on the horizon. An entire market space called "AI Agents" is getting a lot of investment from venture firms and tech companies and will supercharge much more of this integration, as well as evolve the kind of tasks AI can do on our behalf.

iii   As always, proceed with caution. AI can hallucinate, and more subtly, it can overly influence your thinking if not used mindfully.

their viewpoints, surfacing insights that may have been missed and uncovering blind spots. By applying AI as a sort of synthetic perspective that can adopt any role or viewpoint, teams can enhance their analytical capabilities and drive more comprehensive, well-rounded outcomes. For instance, teams can complement human analysis by using AI to identify gaps or missed points in their deliverables, and I've seen teams use this as a standard final check before considering a piece of work complete. AI can also evaluate materials from various stakeholder perspectives, helping teams create more resonant content for diverse audiences. As a "neutral" third party,[iv] AI can assess the pros and cons of different options or surface critical questions in decision-making processes, leading to more thorough evaluations. The applications are vast: from reviewing marketing materials and preparing for negotiations to analyzing new market strategies and product designs.

Skill Filler: AI can fill in for expertise in areas where your team's talent is limited, helping your organization to operate with more agility. Especially as domain-specific tools become more sophisticated, they can provide a range of specialty skills on demand, making it possible to quickly expand your team's capabilities. While AI may not always be capable of completing a project entirely, it can often make significant headway on specialized tasks. These range from language translation and image or video generation to data visualization, coding, social media content creation, technical writing, and SEO optimization. This rapid infusion of skills enables teams to take on more end-to-end projects that were previously out of their scope or

---

iv   No technology is truly neutral, as it reflects the design decisions of its creators and, in AI's case, its training data. However, I've found that people can be more receptive to alternative viewpoints or approaches when presented by AI rather than by another human. This likely stems from several factors. Despite its inherent biases, AI is often viewed as more impartial than humans because it lacks personal agendas or emotional investments. (AI doesn't carry the baggage of past interactions or office politics.) It could also help neutralize egos, as accepting a different viewpoint from AI may feel less like "giving in" to a colleague's opinion. In these early stages, it's possible the novelty of AI plays a role as well, as people may be more curious and open to considering an unexpected AI suggestion than that of a colleague.

to offer a wider range of services.

Talent Multiplier: And finally, when each team member understands how to individually tap into AI for a brain boost, it can elevate the entire team's performance. When used well, each person on your team has on-demand access to a smart, tireless thought partner, ready to jump in when needed to spark fresh ideas and unravel complex challenges. This AI boost can magnify each person's individual strengths, allowing team members to more fully apply their unique talents and expertise to their work—in other words, to be even better at what they do best. While it's not a magic solution, it can be a powerful tool for unlocking more of each person's potential within the team.

Using Custom Chatbots to Support Your Team

Teams can easily create custom chatbots tailored for specific tasks. A great place to start is by experimenting with OpenAI's "GPTs." Additionally, many software providers are beginning to offer the ability to create not only custom chatbots but also AI agents, which are rapidly evolving to manage more sophisticated and complex workflows.

There is a world of inspiration online to see how others have approached this; it can take just a few minutes to set up and doesn't require programming knowledge. It's a great way to experiment with getting value out of the tools, but (as always) start with a clear business need to focus your work.

For example, create a chatbot that analyzes call transcripts to spot patterns and craft customer service responses. Or make a "documentation specialist" that transforms complex product information into user-friendly guides, a meeting bot that identifies action items and assigns tasks, or an "empathy engine" that adopts different customer personas to evaluate your product and marketing strategies through their unique perspectives.

The AI Learning Journey for Teams

I've found a few approaches to be especially helpful to teams, which I've put together in a Team Learning Journey. It works best when team members are also on their own personal AI journeys: individual learning supports the team's collective growth and helps establish a common language across the team. Ideally, use this approach in tandem with the A^3 Framework detailed in Chapters Thirteen through Fifteen.

Consider this chapter only an introduction. Building your team's AI muscle is a substantial undertaking that warrants significant time, attention, and support. When working with organizations, I manage this as an extended process involving multiple workshops, ongoing communication, collaborative discovery of team needs, supported experimentation, and thoughtful change management. However, while the full implementation is beyond the scope of this book, you can use this guidance to begin shaping your approach.

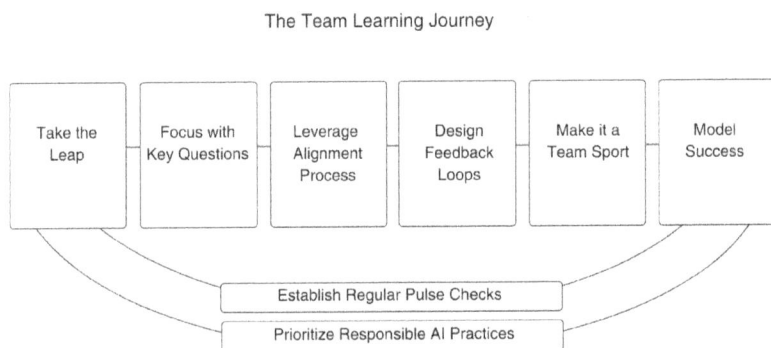

The Team Learning Journey

Take the Leap	Focus with Key Questions	Leverage Alignment Process	Design Feedback Loops	Make it a Team Sport	Model Success

Establish Regular Pulse Checks

Prioritize Responsible AI Practices

Take the Leap: Don't wait for perfect clarity—start your AI journey now. Many teams hesitate, but delay comes with tradeoffs: the opportunity cost of waiting rises every quarter as competitors gain AI experience. And your employees may already be using "shadow AI" without guidance, which means they—and you—are missing out on valuable learning opportunities. Take action now, even if you don't feel fully ready. Waiting for the perfect moment to start may mean never actually beginning.

Focus with Key Questions: Develop and continually refine a set of guiding questions that your team can use as a north star to focus their research and exploration. Be as specific as possible, drilling down to reveal underlying dimensions. A simple but powerful technique is to repeatedly ask "Why is this really important?" for each question. This approach makes for a useful work session on its own. The discussion often uncovers crucial aspects that need exploration: the deeper you dive, the more you'll learn on your journey to find answers. If you're doing this step right, you'll never run out of questions. Instead, you'll constantly evolve existing ones and generate new ones as your understanding grows. Regularly revisit these questions as a team to maintain alignment and adjust your focus as needed.

Leverage the Alignment Process: Adapt Step Two, Align, from Chapter Fourteen to a team context. Begin with a collective assessment of your team's needs. Then, have pairs or small groups choose an AI Challenge to tackle. Encourage regular sharing of discoveries and insights, and promote exchanges among team members to get feedback on their progress. This approach creates a collaborative environment where everyone benefits from the team's expanding collective AIQ.

Design Feedback Loops: The careful design of feedback loops is critical to developing an organization's AI muscle. Thoughtfully cultivate communication channels that will surface concerns and tap into grassroots innovation happening closest to workflows and customers. Use this structure to nurture a shared understanding of responsible AI development and foster meaningful dialogue that supports organizational change. Formalize regular retrospectives or debriefs after implementing new AI approaches, making continuous improvement an integral part of your process.

Make It a Team Sport: Accelerate learning by forming smaller "learning circles," or workgroups of people united by their enthusiasm to explore specific AI topics. Establish a regular format and frequency for sharing sessions where members discuss their discoveries, explore implications, and refine key questions. This approach ensures time is allocated to the open-ended exploration that is so important for driving new insights and can energize learning efforts because people feel accountable to their team members. Additionally, organize team activities that make it fun to explore use cases for AI. I've had clients create "Curiosity Cafes" (informal learning and ideation spaces complete with baristas), "No-Code Hackathons" designed specifically for non-technical newbies, "AI Challenge Leagues" to gamify the exploration process, or "Swap Sessions" where team members can freely discuss AI questions and experiences over food or drinks. Organizations often find there is so much interest in learning about the technology that events are frequently oversubscribed. When this approach works best, it ends up building a community of AI explorers in your organization, and by not only plugging into this community but supporting and celebrating its success, you can foster a great deal of AI momentum in your company.

Model Success with Storytelling: Storytelling is a powerful tool for bringing your organization's AI journey to life. It can vividly demonstrate how AI brings value and supports strategic objectives, while also honestly portraying the challenges encountered along the way—and crucially, how these obstacles were overcome. By finding and sharing these stories more broadly, you can show what success looks like, help to ease fears, and make education on more abstract concepts such as responsible

AI practices, data privacy, and governance more tangible and engaging. Using diverse formats—written case studies, video interviews, or live presentations—helps to cater to different learning styles and encourages team members at all levels to share their experiences.

Establish Regular Pulse Checks: Implement regular checkpoints to assess your team's AI learning progress. Use these sessions to surface new key questions and evaluate the outcomes of your AI experiments. Discuss as a team how to improve these initiatives and identify new areas where AI could add value. Encourage open dialogue to surface team concerns and evaluate your responsible use of AI.

Prioritize Responsible AI Practices: Make discussions about AI ethics a regular part of team meetings to ensure everyone understands and commits to responsible AI practices. If your company has a responsible AI team, collaborate closely with them. If not, advocate for its formation. Integrate ethical considerations into every phase of your work, and regularly assess the ethical implications of your AI use cases.

Navigating AI as a Team

Just as individuals have varying levels of experience with AI, from novice to expert, teams exhibit a wide spectrum of AI proficiency. In the same way it can be challenging for individuals to start thinking with AI, establishing a collective "practice of using AI" within an organization can be difficult. It can be hard to bring all team members to a common understanding of AI capabilities, develop a shared language around AI concepts, and align on where and how to experiment with AI in their work together.

As we discussed in Chapter Six, using AI at the wrong time can hinder our cognitive performance. This principle applies to teams too: we must invest in helping our teams grasp the subtleties needed to prevent AI overreliance from short-circuiting their collective brain power. A recent study found that if teams don't have a deep understanding of when and how to apply AI, it may result in only modest gains in team creativity—and that some AI-using teams even underperformed compared to those relying solely on human intuition.[61]

The teams that truly excel with AI grasp the importance of relying on human cognition during crucial parts of the creative process. When they do turn to AI, they approach it as a collaborative partner, iteratively refining its outputs and exploring creative avenues that might otherwise remain unexplored. It's this nuanced approach—knowing when to lean on human insight and when to tap into AI's capabilities—that enables teams to truly enhance their performance with AI. This underscores the importance of not simply introducing AI into the creative process; organizations must equip their teams with the knowledge and skills to engage effectively.

AI in the Enterprise: Proceed with Care

Leveraging AI in an organization demands careful, proactive consideration of many additional factors. It's not just about more people to coordinate and align on using AI productively; it's also critical to anticipate and understand the broader and downstream impacts of your AI adoption. This means you'll really need to explore as a team how AI can support strategic objectives while managing the new risks and responsibilities it introduces. This involves asking tough questions and considering long-term consequences—and this is hard but critical work. The sooner your team learns to work responsibly with AI, the better prepared you'll be to navigate the inevitable changes it will bring to your industry and market space.

I want to emphasize that the Team Learning Journey shared in this chapter works for a team's journey of learning how to Think with AI. When it comes to developing AI products and services that will impact the people your organization serves—whether internal tools or customer-facing solutions—and taking these to production, you need a completely different, more rigorous process. That process requires extensive testing, evaluation, and compliance with security measures, data handling practices, governance policies, and ethics protocols. As you transition from exploratory learning to developing actual products and services, it's crucial to closely align with your organization's established processes, guidelines, and governance frameworks to ensure responsible and ethical AI development practices.

Speed to Learn over Speed to Launch

There can be great pressure to get AI solutions out the door. But with the space moving so quickly, data infrastructure that still needs work to truly leverage new AI capabilities, and best practices still forming, only the most sophisticated teams have the potential to rush into production without problems, and even these teams have made some very visible and embarrassing missteps. This is a great time to aim to be right to market over first to market. In the Wild West that AI is in these early stages, moving carefully will help you maintain trust with your customers and other stakeholders.

Unless you're an AI company—where competitive pressures demand rapid product launches—I advise my clients to focus on accelerating their learning and increasing the efficiency of that learning process over rushing to bring AI products to market. Prioritizing learning and the deliberate strengthening of AI muscle can lay a strong foundation on which to steadily advance your AI capabilities. And helping your teams to work through the approach outlined in this chapter is a fantastic way to accelerate this learning while simultaneously extracting real business benefit from AI.

Growing Your Team's AIQ: Start Small, Think Big

The unprecedented pace of AI advancement has created a unique moment in time: from entry-level employees to seasoned executives, nearly everyone is simultaneously embarking on this learning journey, regardless of their previous experience with artificial intelligence. This reality presents an extraordinary opportunity: we're all early explorers—and every team has the potential to pioneer something new.

By adopting the learning-centric approach outlined in this book, teams can begin building their AI capabilities no matter how much experience they have. These initial steps are far from trivial. Each interaction, each lesson learned, and each small success builds your team's collective AIQ. This incremental progress can not only enhance immediate team performance but lay a foundation for more ambitious AI initiatives in the future. Every team's AI journey begins with a single step. Those who start now—even with modest initiatives—aren't just learning about today's AI; they're building the collective capability to harness whatever remarkable developments tomorrow brings.

Afterword

Every so often, a technology emerges that fundamentally alters the course of human civilization. Now we face a profound moment: a technology so powerful it blurs the boundaries between human and machine and catapults us into a future where the definition of "possible" is rapidly redrawn.

Like every transformative technology in human history—from the printing press that democratized knowledge, to the steam engine that powered industrialization, to electricity that reshaped daily life, to the internet that connected minds across the globe—AI morphs to the intentions of the hand that wields it.

Every powerful tool carries dual potential: a surgeon's blade can heal or harm, an author's pen can illuminate or obscure truth. Even the internet, while creating the world's greatest library, has also become its most prolific source of misinformation.

Already we see AI following this pattern: accelerating medical breakthroughs while powering sophisticated fraud, helping teachers personalize learning at scale while also scaling the generation of harmful misinformation, expanding human creative potential while threatening the livelihoods of people in long-established careers.

This duality is integral to the story of human innovation. When we release powerful technologies into the world, they become mirrors reflecting the full spectrum of human nature—our highest aspirations and our darkest impulses, our drive to create and our capacity to destroy, even our most mundane struggles and needs. What's different about AI is both its unprecedented power and the staggering speed of its evolution. We're witnessing what may be the fastest technology innovation and adoption cycle in human history, with capabilities advancing not over decades but months.

While we can't control every aspect of AI's development, we can choose how we engage with it. We can make a choice to apply it toward ends that matter—to amplify the positive impact we hope to have.

This is why I'm focused on helping people understand and harness AI's potential to make positive change. Whether you're working to solve global challenges or strengthen your local community, supporting your family or building a business, teaching students or leading a team—your ability to thoughtfully leverage AI can amplify your unique contribution to the world.

I wrote this book to push back against a new digital divide already taking shape, where AI's benefits flow primarily to those who already possess technology advantage. While AI's raw power is available to all, the knowledge of how to wield it effectively is not yet evenly distributed. My hope is that by making this learning journey more accessible, we can help ensure AI's capabilities land in the hands of those working to propel progress in every field and community.

The path ahead isn't entirely clear. We'll face difficult questions about how to better align AI with human values. We'll need to work through thorny challenges around equity, accountability, and control. There will be setbacks along the way. But I remain deeply optimistic that the more who understand this technology, the better equipped we will be as a society to shape it into a force that drives human flourishing—not with blind techno-optimism, but with clear-eyed determination to make it so.

If you've found value in these pages, I hope you'll share these tools and insights with others who could use them to amplify their own positive impact. I hope you'll join the growing community of thoughtful practitioners working to develop effective and responsible ways to harness AI's capabilities. Most of all, I hope you'll bring your unique perspective and wisdom to help shape how this technology evolves.

The story of AI will be written by millions of people making millions of choices about how to use these new capabilities. Your voice, your choices, your discoveries matter more than you might think. I invite you to stay connected through ThinkwithAI.org and LinkedIn, where I will continue to share insights that help us all use this technology more effectively.

We stand at the beginning of a new chapter in human capability. The possibilities ahead are both exhilarating and sobering. But I've never been more convinced that by thoughtfully engaging with AI—by learning to truly think with these new tools—we can expand our human ability to solve problems, create value, and advance humanity in ways we're only beginning to imagine.

Let's write this next chapter together.

Alison McCauley is a digital transformation strategist dedicated to advancing human potential. Working at the nexus of the enterprise and disruptive innovation, she helps executives navigate tech-driven change. Alison works with leaders across the ecosystem, coaching organizations to amplify performance by harnessing emerging technology while helping tech pioneers to inspire their stakeholders to take action.

She began her career at the leading edge of change management as the field was first coalescing into a formal discipline—a transformative period that revealed how profoundly human behavior and culture shape technology adoption. This early immersion became the foundation of her thirty-year focus on helping organizations to advance performance, including a decade working with AI pioneers.

A best-selling author whose insights appear in *Harvard Business Review* and *Forbes*, Alison is an in-demand keynote speaker at conferences worldwide and her executive education course on LinkedIn has reached over 100,000 learners.

Alison's commitment to making emerging technology accessible to all continues in her current role as Chief Advocacy Officer of Think with AI, where she empowers people—regardless of their experience or background—to enhance their unique capabilities through AI. She is also founder of Unblocked Future, a strategy firm focused on accelerating the effective adoption of transformative technology and unlocking the business value it creates.

She is deeply committed to how technology can advance the future of work, create more sustainable cities, enhance human health and wellness, and strengthen meaningful connection in our digital world. She holds a BA in Psychology and an MA in Organization Design and Development, both from Stanford University.

Notes

1   Ilya Sutskever, "Rogue superintelligence and merging with machines: Inside the mind of OpenAI's chief scientist," MIT Technology Review, October 26, 2023, https://www.technologyreview. com/2023/10/26/1082398/exclusive-ilya-sutskever-openais-chief-scientist-on-his-hopes-and-fears-for-the-future-of-ai/.

2   Mustafa Suleyman, "What is an AI anyway?" [Video], TED Conferences, April 2024, https://www.ted.com/talks/mustafa_suleyman_what_is_an_ai_anyway.

3   David Brooks, "A.I.'s Benefits Outweigh the Risks," New York Times, July 31, 2024, https://www.nytimes.com/interactive/2024/07/31/opinion/ai-fears.html.

4   Vijay Balasubramanian, "Brain Power," Proceedings of the National Academy of Sciences 118, no. 32, August 10, 2021, https://www.ncbi.nlm.nih.gov/pmc/articles/PMC8364152/.

5   N. M. Daumeyer, I. N. Onyeador, X. Brown, and J. A. Richeson, "Implicit-Bias Remedies: Treating Discriminatory Bias as a Public Health Problem," Policy Insights from the Behavioral and Brain Sciences, 2019, https://www.ncbi.nlm.nih.gov/pmc/articles/PMC9121529/.

6   Miguel Fernández-Armesto, "Cognitive Short-Sightedness, Biases and Survival," CCCB LAB, Centre de Cultura Contemporània de Barcelona, February 8, 2019, https://lab.cccb.org/en/cognitive-short-sightedness-biases-and-survival/.

7   R. Hanson, "How Your Thinking Affects Your Brain Chemistry," Psychology Today, January 31, 2023, https://www.psychologytoday.com/us/blog/the-courage-happiness/202301/how-your-thinking-affects-your-brain-chemistry.

8 J. Taylor, "Neither Fight nor Flight Helps Us 'Survive' in Modern Times," Psychology Today, March 14, 2022, https://www.psychologytoday.com/us/blog/the-power-prime/202203/neither-fight-nor-flight-helps-us-survive-in-modern-times.

9 T. H. Harvard Chan School of Public Health, "Stress and Health," The Nutrition Source, October 2021, https://nutritionsource.hsph.harvard.edu/stress-and-health/.

10 National Institute of Mental Health, Attention-Deficit/Hyperactivity Disorder (ADHD), https://www.nimh.nih.gov/health/statistics/attention-deficit-hyperactivity-disorder-adhd.

11 E. Kerr, "What Is Dyslexia?" U.S. News & World Report, April 17, 2023, https://www.usnews.com/education/k12/articles/what-is-dyslexia.

12 Michael Ignatieff, "Epistemological Panic, or Thinking for Yourself," Liberties 3, no. 2, Winter 2023.

13 A. Humlum and E. Vestergaard, "The Adoption of ChatGPT," July 9, 2024, https://static1.squarespace.com/static/5d35e72fcff15f0001b-48fc2/t/668d08608a0d4574b039bdea/1720518756159/chatgpt-full.pdf.

14 Google. "This is changing the way scientists research | Gemini," YouTube video, 10:15, https://www.youtube.com/watch?v=sPiOP_CB54A.

15 Fabrizio Dell'Acqua, "Falling Asleep at the Wheel: Human/AI Collaboration in a Field Experiment on HR Recruiters," Working Paper, Laboratory for Innovation Science, Harvard Business School, July 2022, https://static1.squarespace.com/static/604b23e38c22a96e-9c78879e/t/62d5d9448d061f7327e8a7e7/1658181956291/Falling+Asleep+at+the+Wheel+-+Fabrizio+DellAcqua.pdf.

16 U. Agudo, K. G. Liberal, M. Arrese, et al., "The Impact of AI Errors in a Human-in-the-Loop Process," Cognitive Research: Principles and Implications 9, no. 1, January 2, 2024, https://doi.org/10.1186/s41235-023-00529-3.

17   David Brooks, "A.I.'s Benefits Outweigh the Risks," New York Times, July 31, 2024, https://www.nytimes.com/interactive/2024/07/31/opinion/ai-fears.html.

18   Brené Brown, "The Call to Courage," directed by Sandra Restrepo, Netflix, April 19, 2019, https://www.netflix.com/title/81010166.

19   Zher-Wen and Rongjun Yu, "Perceptual and semantic same-different processing under subliminal conditions," Consciousness and Cognition, 111, 2023, https://doi.org/10.1111/bjop.12631.

20   C. Soon, M. Brass, H. J, Heinze, et al., "Unconscious determinants of free decisions in the human brain," Nature Neuroscience 11, 543–545, 2008, https://doi.org/10.1038/nn.2112.

21   M. Beenman and J. Kounios, "The Aha! Moment: The Cognitive Neuroscience of Insight," Current Directions in Psychological Science 18, no. 4, August 2009, https://doi.org/10.1111/j.1467-8721.2009.01638.x.

22   David Eagleman, Incognito: The Secret Lives of the Brain, Pantheon Books, 2011.

23   S. M. Ritter, R. B. van Baaren, and A. Dijksterhuis, "Creativity: The Role of Unconscious Processes in Idea Generation and Idea Selection," Thinking Skills and Creativity 7, no. 1, 2012, 21-27, https://doi.org/10.1016/j.tsc.2011.12.002.

24   A. Dijksterhuis and T. Meurs, "Where Creativity Resides: The Generative Power of Unconscious Thought." Consciousness and Cognition 15, no. 1, March 2006, 135-146, https://doi.org/10.1016/j.concog.2005.04.007.

25   T. K. Gandhi, D. Classen, C. A. Sinsky, et al., "How can artificial intelligence decrease cognitive and work burden for front line practitioners?" JAMIA Open, August 29, 2023, https://doi.org/10.1093/jamiaopen/ooad079.

26   L. Bai, X. Liu, and J. Su, "ChatGPT: The Cognitive Effects on Learning and Memory," Brain and Behavior 13, no. 11, November 2023, https://doi.org/10.1002/brx2.30.

27   Heinrich von Kleist, "On the Gradual Construction of Thoughts During Speech," trans. Michael Hamburger, German Life and Letters 5, no. 1, 1951, 42-46, https://www.ias-research.net/wp-content/uploads/2012/01/Kleist-and-Hamburger_-_1951_-_On-the-Gradual-Construction-of-Thoughts-During-Speech.pdf.

28   E. Xu, W. Wang, and Q. Wang, "The effectiveness of collaborative problem solving in promoting students' critical thinking: A meta-analysis based on empirical literature," Humanities and Social Sciences Communications 10, 16, 2023, https://doi.org/10.1057/s41599-023-01508-1.

29   Roger E. Beaty and Yoed N. Kenett, "Associative Thinking at the Core of Creativity," Trends in Cognitive Sciences 27, no. 7, July 2023, 671-683, https://doi.org/10.1016/j.tics.2023.04.004.

30   Ethan Mollick, Co-Intelligence: Living and Working with AI, 2024, Portfolio/Penguin Random House.

31   Adam Cheyer, "What is the Case for Techno-Optimism Around AI?" Talk presented at Reinvent Futures, The AI Age Begins, 2024.

32   OpenAI, GPT-4 Technical Report, 2023 https://arxiv.org/pdf/2303.08774.

33   E. E. Guzik, C. Byrge, and C. Gilde, "The Originality of Machines: AI Takes the Torrance Test," Artificial Intelligence in Medical Sciences, 4, 2023, https://doi.org/10.1016/j.yjoc.2023.100065.

34   B. Lev and T. Sougiannis, "Ideas are Dimes a Dozen: Large Language Models for Idea Generation in Innovation," Social Science Research Network, 2023, https://doi.org/10.2139/ssrn.4526071.

35   F. Salvi, M. H. Ribeiro, R. Gallotti, and R. West, "On the Conversational Persuasiveness of Large Language Models: A Randomized Controlled Trial," 2024, https://doi.org/10.48550/arXiv.2403.14380.

36   A. Karthikesalingam and V. Natarajan, "AMIE: A research AI system for diagnostic medical reasoning and conversations," Google Research Blog, January 12, 2024, https://research.google/blog/amie-a-research-ai-system-for-diagnostic-medical-reasoning-and-conversations/.

37   J. W. Ayers, A. Poliak, M. Dredze, et al, "Comparing Physician and Artificial Intelligence Chatbot Responses to Patient Questions Posted to a Public Social Media Forum," JAMA Internal Medicine 183, 589–596, https://jamanetwork.com/journals/jamainternalmedicine/fullarticle/2804309.

38   Ethan Mollick, "Everyone is Above Average," One Useful Thing, September 24, 2023.

39   Fabrizio Dell'Acqua, Edward McFowland III, Ethan R. Mollick, et al., "Navigating the Jagged Technological Frontier: Field Experimental Evidence of the Effects of AI on Knowledge Worker Productivity and Quality," Social Science Research Network, September 18, 2023, https://papers.ssrn.com/sol3/papers.cfm?abstract_id=4573321.

40   Shakked Noy and Whitney Zhang, "Experimental Evidence on the Effects of Generative Artificial Intelligence," March 10, 2023, https://economics.mit.edu/sites/default/files/inline-files/Noy_Zhang_1_0.pdf.

41   Fabrizio Dell'Acqua, Edward McFowland III, Ethan R. Mollick, et al., "Navigating the Jagged Technological Frontier: Field Experimental Evidence of the Effects of AI on Knowledge Worker Productivity and Quality," Social Science Research Network, September 18. 2023, https://papers.ssrn.com/sol3/papers.cfm?abstract_id=4573321.

42   Erik Brynjolfsson, Danielle Li, and Lindsey Raymond, "Generative AI at Work," National Bureau of Economic Research, October 9, 2023, https://doi.org/10.48550/arXiv.2304.11771.

43   David Autor, Caroline Chin, Anna Salomons, and Bryan Seegmiller, "New Frontiers: The Origins and Content of New Work, 1940–2018," National Bureau of Economic Research Working Paper, August 2022, https://www.nber.org/papers/w30389.

44   "Global influencer marketing market size from 2016 to 2022 (in billion U.S. dollars)," Statista.

45   Matt Levine, "Marc Andreessen Answers the Tech Valuation Question," Bloomberg, May 22, 2017, https://www.bloomberg.com/

opinion/articles/2017-05-22/marc-andreessen-answers-the-tech-valuation-question.

46   Nick Bilton, "Bubble? What Bubble?" New York Times, July 10, 2011, https://www.nytimes.com/2011/07/10/magazine/marc-andreessen-on-the-dot-com-bubble.html.

47   Clifford Stoll, "Why the Web Won't Be Nirvana," Newsweek, February 26, 1995, https://www.newsweek.com/clifford-stoll-why-web-wont-be-nirvana-185306.

48   Alexander Bick, Adam Blandin, and David Deming, "The Rapid Adoption of Generative AI," Federal Reserve Bank of St. Louis Working Paper 2024-027, September 2024, https://doi.org/10.20955/wp.2024.027.

49   Anson Ho, Tamay Besiroglu, Ege Erdil, et al., "Algorithmic progress in language models," 2024, https://doi.org/10.48550/arXiv.2403.05812.

50   Ashish Vaswani, Noam Shazeer, Niki Parmar, et al., "Attention is all you need," 2017, https://doi.org/10.48550/arXiv.1706.03762.

51   Benedict Evans, "The AI summer," July 9, 2024, https://www.ben-evans.com/benedictevans/2024/7/9/the-ai-summer.

52   Mustafa Suleyman, "What is an AI anyway?" [Video], TED Conferences, April 2024, https://www.ted.com/talks/mustafa_suleyman_what_is_an_ai_anyway.

53   Ethan Mollick, "Superhuman?" One Useful Thing, May 7, 2024, https://www.oneusefulthing.org/p/superhuman.

54   E. M. Woodin, "A two-dimensional approach to relationship conflict: Meta-analytic findings," Journal of Family Psychology, 325-335, 2011, https://doi.org/10.1037/a0023791.

55   Cheng Li, Jindong Wang, Yixuan Zhang, et al., "Large Language Models Understand and Can be Enhanced by Emotional Stimuli," 2023, https://doi.org/10.48550/arXiv.2307.11760.

56   Alex Tamkin, Amanda Askell, Liane Lovitt, et al., "Evaluating and Mitigating Discrimination in Language Model Decisions," December 6, 2023, https://arxiv.org/pdf/2312.03689.pdf.

57   "Learning through Play: What the Science Says," The LEGO Foundation, https://learningthroughplay.com/explore-the-research/the-scientific-case-for-learning-through-play.

58   "The Power of Play," The Hechinger Report, https://hechingerreport.org/the-power-of-play/.

59   Amy Jo Dowd, "In Times of Uncertainty, Our Most Creative Thoughts Can Happen Through Play," Fast Company, December 16, 2020, https://www.fastcompany.com/90593529/in-times-of-uncertainty-our-most-creative-thoughts-can-happen-through-play.

60   James Clear, Atomic Habits: An Easy & Proven Way to Build Good Habits & Break Bad Ones, Avery, 2018.

61   "Don't Let Gen AI Limit Your Team's Creativity," Harvard Business Review, March 2024, https://hbr.org/2024/03/dont-let-gen-ai-limit-your-teams-creativity.